GENES, CELLS, AND BEHAVIOR

A Series of Books in Biology

Cedric I. Davern, Editor

GENES, CELLS, AND BEHAVIOR:

A View of Biology
Fifty Years Later

50th Anniversary
Symposium

Division of Biology
California Institute of Technology

Edited by **Norman H. Horowitz**
and **Edward Hutchings, Jr.**

November 1-3, 1978

W. H. Freeman and Company
San Francisco

Cover art from Sinsheimer, page 47.

Library of Congress Cataloging in Publication Data

Main entry under title:

Genes, cells, and behavior.

(A Series of books in biology)
Includes bibliographies and indexes.
1. Biology—Congresses. 2. Cancer—Congresses.
3. Psychobiology—Congresses. I. Horowitz, Norman
Harold, 1915- II. Hutchings, Edward.
III. California Institute of Technology, Pasadena.
Division of Biology. [DNLM: 1. Biology—
Congresses. QH307.2 G327 1978]
QH301.G39 574 80-18744
ISBN 0-7167-1217-2

Printed in the United States of America

1 2 3 4 5 6 7 8 9

Preface

The papers in this book were originally presented in a symposium of the same title, held in November, 1978, on the campus of the California Institute of Technology. The occasion was the celebration of the 50th Anniversary of the founding of the Division of Biology at Caltech. The Division was formed in 1928 by Thomas Hunt Morgan, its first chairman. The original faculty consisted entirely of geneticists: A. H. Sturtevant, E. G. Anderson, Sterling Emerson, and Morgan. Calvin Bridges was a research associate, and Theodosius Dobzhansky was a research fellow. With such a beginning, it is not surprising that genetics in its various manifestations--classical, microbial, biochemical, and now molecular--has remained central to the interests of the Division, although cell biology and, especially, neuroscience have also become important disciplines at Caltech in recent years.

For the Jubilee celebration, we desired to have a reunion and also a scientific event worthy of the occasion. To these ends, it was decided to organize a symposium on current problems in biology in which all the speakers would be alumni or former members of the Division. The result is presented in this book. The reader will find here a collection of overviews of their current research by a group of notable biologists, written for a general biological audience. My only regret is that it was impossible within the limits of a $2\frac{1}{2}$-day symposium to include in the program a larger number of our distinguished alumni, many of whom were in the audience.

The order of papers as presented here follows the sequence of the symposium. One paper, that of David Hogness, was unavailable. The chairmen's introductions were somewhat modified for publication, but Delbrück's account of the history of phage research at Caltech is essentially as it was presented. Eventually, as the river of science flows on, this book will go out of date, but it will remain an historical document; perhaps it will be quoted at the Centennial celebration in 2028 to show what some of us were thinking and doing in the 1970's. I hope so.

I wish to express my thanks to the authors and chairmen for their cooperation in making this publication possible. Thanks are due also to Connie Katz and Bernita Larsh for typing the manuscript, to Geraldine Cranmer and Elizabeth Koster for proofreading it, and to David Asai, Larry Johnson, and Donna Livant for preparing the Indexes.

N. H. Horowitz, Chairman
Division of Biology
California Institute of Technology

Financial sponsors of the Symposium were:

National Science Foundation
Calbiochem-Behring Corporation
New England BioLabs, Inc.
Boehringer Mannheim Biochemicals
Pierce Chemical Company
McBain Instruments, Inc.
New England Nuclear
N. H. Horowitz
Division of Biology, California Institute of Technology

Table of Contents

GENES, CELLS, AND BEHAVIOR

Introduction

R. D. Owen

Professor of Biology, Caltech

In celebrating 50 years of biology at Caltech, it may seem capricious to begin with selections from contributions of Divisional people to "cancer research." For the most part, the Division has prided itself through its history on its devotion to basic experimental biology, not to the solution of practical human problems, however pressing. The general impression has been fostered that meeting the intellectual challenges of understanding life and its processes at primary levels will lead to eventual control of processes at all levels, including those from abnormal cellular behavior to organismal disease. The implication has been that "applied" developments are to be left to others elsewhere, not to distract us from the "basic" research we have tried to do well and have thought it best for us to do.

The four papers in this section represent a step toward the "applied" against this background, because they deal with a compelling human problem, cancer. In different ways, they illustrate the power of the principles upon which the Biology Division at Caltech has been built. When Renato Dulbecco came to Caltech as a Senior Research Fellow in 1949, to join Delbrück's group after an interval with Luria, high excitement characterized the use of bacteriophage as a tool for understanding primary attributes of life. There was at the time, I am sure, in the back of many minds an idea that what was being found out about viral infections of E. coli would relate, in time, to practically important matters of plant and animal viral disease. But that set of possibilities was not evident as a prime motivation for the research; the deepening insight, being able to put biological phenomena into a quantitative framework of understanding, was immediate reward enough. And when Dulbecco turned to the animal viruses on the basis of the phage experience, including the very provocative aspects of the incorporation of lysogenic phages into the host cell's genome, it was with a sense of extension of that same kind of excitement. First with polio viruses, and adapting with modification the quantitative methods of phage research using animal cells in culture and derivatives of plaque-counting techniques, and then with tumor viruses, Dulbecco established a laboratory at Caltech in the forefront of one very important aspect of cancer research. His move to the Salk Institute in 1963 and later to become a Deputy Director of The Imperial Cancer Research Fund in London, now back again at the Salk Institute and a Professor of Pathology at the University of California, San Diego, Medical School, was a loss to Caltech but not to Science.

In the first paper in this section, he gives an up-to-date account of the incorporation of the DNA of a class of animal tumor viruses into the genomes of their host cells, and of subsequent processes of transformation of these cells into cancer cells. He discusses evidence, in turn, that the viral genes themselves are ultimately of cellular origin, genes that were initially controlled by repressor mechanisms similar to those known in bacteria, but escape from repression, possibly by mutation in the controlling genes. The hypothesis provides a basis for understanding the provocative relationship between spontaneous or induced cancers and viral transformation.

If Dulbecco's paper can be described as a child of the phage group, Temin's is a grandchild. Three generations of Nobel Prize work--Delbrück to Dulbecco

1

to Temin! Temin's paper deals primarily with a kind of virus that was not envisioned in the early phage work--tumor viruses whose primary genetic material is RNA rather than DNA. Temin was, of course, a graduate student in Dulbecco's group, who came to Caltech after taking his Bachelor's degree at Swarthmore in 1955. Four years later, with a Caltech Ph.D., he joined the staff of the McArdle Laboratory for Cancer Research at the University of Wisconsin, where he is currently an American Cancer Society Professor of Viral Oncology and Cell Biology. I remember Temin's fascination, as a graduate student, with the RNA viruses and his speculations about how they might operate. In his paper, he describes how the "retroviruses," RNA viruses including the rapidly oncogenic Rous sarcoma virus, make DNA copies of their RNA, which become inserted in the DNA of the host cells. He deals with his hypothesis, for which there is substantial evidence, that the "weak" viruses, which are relatively slow to transform cells, evolved from genes in the host species, and he describes how such viruses may develop, evolve, and escape. Their ability to enter cells as RNA, to integrate into the host genome as DNA, and to come out again as RNA has provided not only an intellectually fascinating subject for investigation, but is also conceptually an important part of the revolution in genetics associated with the control of recombinant DNA. No doubt the most important potential contribution from this kind of research is still to come--an understanding not only of the abnormal growth we call cancer but of some of the processes of normal differentiation and development as well.

Although the environment the Caltech Biology Division has provided its students has always included easy interactions among people with various interests, it is fair to say that Bruce Ames' immediate surroundings were considerably different from those in the virology group, when Ames was a graduate student here. He came to work with Herschel Mitchell in 1950, after having taken his Bachelor's degree at Cornell University, and he received his Caltech Ph.D. in 1953. Then he went to the National Institutes of Health, returning to California, the University of California at Berkeley, as a Professor of Biochemistry in 1968. As a student in Mitchell's laboratory, his intellectual lineage therefore traces through Mitchell to Beadle, and the active traditions of biochemical genetics, especially of Neurospora. Mitchell's style with graduate students is relatively permissive; he is always there, attentive and helpful, but he has not chosen to restrict the problems his students might undertake to a narrow and intense range of topics in which he was himself most immediately active at the time. I suspect that this tolerance for diversity is related to how well Mitchell's students have, on the whole, done so many different things when they left Caltech for the outside world. Ames' primary research interests over the years have been closely relevant to those of traditional biochemical genetics--histidine biosynthesis, the regulation of metabolism and protein synthesis, and bacterial biochemical genetics. His paper, however, describes an excursion from this frontier, into territory most directly and importantly relevant to human health, the identification of environmental carcinogens. His expertise in bacterial biochemical genetics led him to develop laboratory strains of Salmonella sensitive to mutagens, and most importantly, able to distinguish a variety of mechanisms for mutagenesis. The strong overlap between carcinogenesis and mutagenesis tempted him to apply these quantitative techniques to the identification of carcinogens. And because important classes of carcinogens in man are not themselves carcinogenic in the form in which they exist in the environment, but are metabolized into effective carcinogens in the body, Ames modified his techniques to identify such classes of carcinogens as well. His paper is remarkable for its intellectual quality, its scope and relevance. Perhaps the greatest challenges it develops are in the social sciences; recognizing the seriousness of the problems created by environmental carcinogens marketed and released in industrial society, what, practically, can or should we do about them?

2

Knudson's, the final paper in this section, illustrates still another kind of Caltech lineage. Like Ames and Temin, Knudson was a graduate student at Caltech in the 1950's. His immediate associations were with the biochemistry of Henry Borsook, and his extensions have been identifiable with the genetics of Morgan, Sturtevant and Lewis. He had received a Bachelor's degree at Caltech in 1944, but then he went to the College of Physicians and Surgeons at Columbia for an M.D. degree, went into pediatrics, and came back to Caltech to take his Ph.D. with Borsook in 1956. He went from Caltech to the City of Hope National Medical Center, where he was Chairman of the Department of Pediatrics and later the Department of Biology; he moved to the New York University system at Stony Brook to become Professor of Pediatrics and Associate Dean of the Health Science Center as the new medical school was developing there, and then to the University of Texas in Houston where he started a biomedical graduate school and served as Professor of Medical Genetics and Dean. Relatively recently, he moved East again to become Director of the Institute for Cancer Research in Philadelphia. His paper deals with heredity and cancer in man. It should serve to clarify an area still subject to considerable confusion among cancer research workers generally--the subject, in fact, of this entire section, Genetics and Cancer. Is cancer "inherited"? We are reminded of the old and, in present light, essentially senseless arguments on nature and nurture. A very large proportion of human cancers are induced by environmental agents--irradiation, chemicals and viruses--and are therefore in principle preventable. But there is good reason to believe that these agents act primarily through their effects on the genetic materials and apparatus, and on hosts that vary in their genetic susceptibility. Knudson's clear delineation of the relationship between genetics and cancer caps this section of the Conference in a most fitting way. And, with the other papers, it reinforces Dulbecco's concluding words: "On this 50th Anniversary we can contemplate with some satisfaction the progress made in the field of cell transformation and cancer. As in many other fields of biology, the experiments were initiated in these laboratories and then spread to many other places, forming a community of effort that looks to this Division as its alma mater."

CELL TRANSFORMATION AND CANCER

Renato Dulbecco, Nobel Laureate

The Salk Institute, La Jolla

When I came here about 30 years ago from Bloomington, Indiana, I was greeted by Max Delbrück. I was going to work with him and this, of course, was a very significant event for me because I think Max is the father of all of us who have been virologists at one time or another and we are really very much indebted to him.

Here I started working with animal viruses. Actually, I think I was the first animal virologist at Caltech, and then Marguerite Vogt came to work with me soon afterwards. We spent several years working with cell-killing viruses like polio virus, and then we turned to oncogenic viruses which don't kill the cells but alter them, making them similar to cancer cells. In fact, these viruses do cause cancer in animals.

In my laboratory I had already had some work done in this field, first by Harry Rubin and then by Howard Temin. And they already had made a mark in the field. They were working with a virus with a very famous name, the Rous sarcoma virus. A few years later, around 1960 I would say, I got involved with another oncogenic virus which had just been discovered and that was the polyoma virus; later I also worked with SV40, which is another related virus. I am going to talk to you especially about this type of work and the consequences I can see out of this work.

These viruses, of course, are quite different from the Rous sarcoma virus that Harry and Howard were using. For instance, they have a double-stranded cyclic DNA as genome, whereas the Rous virus has RNA. However, Howard soon found out that within the cells all these viruses do quite the same things, and the outcome in both cases is that cells are changed in a characteristic way which most people call transformation. But Howard always called it conversion.

These names, of course, reveal our background in microbiology--our way of thinking, which was microbiological. So transformation or conversion, if you will, is the emergence of a cell lineage in which all cells are characteristically altered. But let's look a moment at what the alterations are. In the early 60's it was easy to define the transformed cells as cells that grow well with a little serum, that have unusual morphology.

These changes defined transformation operationally and allowed us to work. But in time, the number of characteristics of transformed cells has grown immensely. These changes affect different properties of the cells, in the sense that different cell compartments seem to be involved at the same time. This raised an important question: How is it possible that by introducing a virus we can change the cell in such a fantastically complicated way?

One can introduce a certain kind of order and rationalization by recognizing that some of the changes of transformed cells are probably produced by a common alteration. For instance, we can think that a change in the cytoskeleton will alter the agglutinability; so these two characters go together. Also, a change in the adhesion to the substrate may change the morphology of the cells.

Another kind of clarifying notion in this complexity is that the properties that are present in a given transformed cell are influenced by certain recognizable factors. For instance, one factor will be the cell type we start with before infection. Another factor will be the transforming virus. Another factor will be the procedure by which the transformed cell is isolated, whether it is isolated by looking for different colonial morphology on plastic or by plating in agar. These and other observations suggest another rationalization,

4

speculative as it may be, i.e., that the multiple differences of transformed cells are produced not only by the action of the virus but also by the state of cellular genes in the particular transformed cell that we examine.

Some differences between the cells may be due to the different states of differentiation of the cells before infection with the virus. Other differences may be introduced subsequently by mutation. Cells with certain changes can be selected by the culture conditions and the media in which we maintain these cells. So it is inevitable that the complex phenotypes of transformed cells result from the summation of events--both viral and cellular.

However, even with the rationalization--that a lot of the complexity of the transformed phenotype is due to the different backgrounds of cellular genes--we must accept the fact that a given transformed cell always has many types of changes that are independent of each other and cannot be easily explained. The action with the virus seems to be pleiotropic. The elucidation of this pleiotropism is actually a major challenge in the study of transformation. As we shall see, it is also a very important clue for understanding the mechanism for transformation.

When we started the work on transformation, we had as a model, or as a hypothesis, that cell transformation must be the counterpart of bacterial lysogeny, namely that it reflects the establishment of a permanent association between a viral and a cellular genome with consequent changes of the cellular phenotype. And our experiments were guided by this model. The immediate success of this approach was the demonstration that the DNA of polyoma virus becomes integrated in the cellular DNA when transformation occurs. This result was very useful because it tended to eliminate one of the hypotheses that were around at the time about transformation, what people used to refer to as the "hit and run" hypothesis. According to this view, the virus hit the cells and then disappeared. This idea was based on the observation that cells transformed by polyoma or SV40 lack the classical evidences of viral infection, or of viral multiplication. However, the DNA persists, and we could also show that a segment of this DNA is transcribed and specific viral proteins are made. These findings then suggested a different hypothesis: that cellular transformation is the consequence of the expression of one or more viral genes superimposed onto the background of cellular genes.

This molecular work was supported and complemented by genetic work which started out in a number of laboratories both with the Rous virus and with the polyoma viruses. The main aim of this work was to find viral mutations affecting transformation. In effect, certain mutants were found. The most striking were temperature-sensitive mutants of the Rous sarcoma virus that affect transformation. The state of the cells transformed by these mutants is entirely dependent on the incubation temperature. At low temperature, the cells have all the characteristics of transformation, but when they are shifted to a higher temperature, they revert to normality. The reversion occurs quite rapidly and affects many cellular characteristics, such as morphology, arrangement of the cytoskeleton, and mobility of surface proteins.

These results allowed the identification of the transforming gene of the Rous virus, which is now called the sarc gene. With polyoma virus, two types of mutations were found to affect transformation. The first one, in the A gene, was isolated here by Mike Fried. This mutant is temperature-sensitive, and at first sight resembles the sarc mutants because it transforms cells only at low temperature and not at high temperature. However, the A mutants are unlike the sarc mutants because the transformed cells, once they are generated at low temperature, tend to remain transformed when they are shifted to high temperatures. Therefore, this A gene of polyoma virus and the sarc gene of the Rous virus have clearly different roles in transformation. The sarc gene controls the maintenance of transformation whereas the A gene of polyoma virus mainly controls events that occur once when the permanent lineage of

5

transformed cells is produced. At least this is the most evident role for this gene. In addition, under certain conditions, perhaps related to the background of functional cellular genes, gene A contributes to the maintenance of transformation, possibly only for certain characteristics.

The mutants in the gene A of polyoma virus can still express transforming function at the nonpermissive temperature because they can induce an abortive transformation. In this type of transformation, the alterations last in the cell lineage for maybe five or six generations and then the cells revert to the normal state. Evidently, the abortive transformation does not require the essential initial event, performed by the A gene, that is required for generating a permanently transformed cell lineage. The event required for permanent, but not for abortive transformation, may be the integration of the viral DNA in the cellular DNA. This is not completely established. It is plausible that the A gene has an integration function because its product interacts with the DNA. However, the main function of this gene is to initiate the autonomous replication of the cyclic viral DNA in the lytic infection.

These studies with the A gene show that the maintenance of transformation by polyoma virus must be due to another gene. Indeed, a second type of mutation was subsequently isolated by Tom Benjamin, who is also a Caltech alumnus. He isolated HRT mutants, which are so called because they have an altered host range and do not transform. These mutations are either small deletions or point mutations. They abolish the ability of the virus to cause either permanent or abortive transformation; therefore, they completely suppress the transforming ability of polyoma virus. Small deletions in the corresponding area of the map of SV40 produced in vitro by beautiful technology in Paul Berg's laboratory at Stanford have similar properties.

These deletions also affect the maintenance of transformation but in a somewhat different way. They cause an incomplete transformation which has fewer and less pronounced changes than regular transformation. It is not very clear why transformation is incomplete. Probably the difference between the polyoma and SV40 mutations can be understood in terms of the interrelation of the expression of viral genes with the cellular genetic background because the cells that are used in the two types of experiments are different.

Once the viral genes involved in transformation were identified, the emphasis shifted to the proteins specified by the genes. With the DNA viruses, this task was facilitated at the beginning by an older discovery of Black and Huebner. They found that the serum of animals carrying a tumor induced by polyoma virus or by SV40 reacts with proteins present in cells infected or transformed by the corresponding virus. These virus-specific proteins are collectively known now as T antigen--tumor antigen. Using this antiserum, the T antigen and the constituent proteins could be purified by immunoprecipitation and gel electrophoresis.

The results show that there is a large T, of molecular weight 90,000 to 100,000. The size of the molecule is affected by temperature-sensitive mutations in gene A and is made smaller by its deletions; therefore, the large T is the product of the A gene. Another protein is the small T (molecular weight ca. 20,000 daltons); this protein is made smaller by deletions in the hr-t gene; therefore, it is the product of that gene. A third protein is the middle T (molecular weight ca. 55,000 daltons), so far only recognized with polyoma virus. This protein is lacking in cells infected by some hr-t mutants.

It is interesting to look at the relationship between these T proteins. They have a common amino acid sequence in the amino end; therefore, they have a common initiation. In fact they are specified by the same DNA segment. How this happens has been shown by the work of several laboratories which have brilliantly revealed the interesting molecular mechanisms. The possibilities of sequencing both the viral DNA and the viral proteins were especially useful in this work. Also useful was the advanced state of the art for translating specific messenger RNAs in vitro. In brief, it was found that the small T is specified by the primary transcript of the transforming region of the viral genome. The protein is small because translation is interrupted by a termination signal. The large T is obtained from the same transcript after splicing, which leads to excision of the termination signal. The site of the hr-t mutations is eliminated at the same time. The middle T is obtained from the same primary transcript, after a different splicing which again removes the termination signal; but this splicing also changes the reading frame for translation.

As to the functions of the three T proteins, our knowledge is still limited. The large T, which is present in the cell nucleus, is a DNA binding protein; it is required for the replication of the viral DNA, which occurs in the nucleus, and as I already said, possibly also for the integration of the viral DNA in transformation. The middle T is present in the cellular plasma membrane; we may speculate that it affects the function of the membrane in growth regulation. The function of the small T remains not understood. From the study of mutants, it seems that all three T proteins are involved in transformation; but the most important role seems to be that of the small T which expresses the hr-t gene.

These transforming proteins may seem too much; but in fact they help to explain the pleiotropism of transformation because we have three different functions. This works only in part because the changes of the transformed cells cannot be all explained on the basis of three primary effects. A recent discovery of Ray Erikson with the Rous virus offers a new, very interesting possible explanation of the pleiotropism. Erikson first produced the transforming protein of the Rous virus by translating in vitro an appropriate viral messenger RNA. This protein is the expression of the sarc gene. The same protein is present in the virus-infected cells. But the most interesting thing is that this protein is a protein kinase, or maybe a protein that regulates a cellular kinase.

The experiments were done in the following way. Ray Erikson made an antiserum against the transformed cells with which he precipitated an extract from the transformed cells. After adding [^{32}P]ATP to this immunoprecipitate, he found that the immunoglobulins became highly labeled with the ^{32}P label: a kinase activity in the immunoprecipitate labels the heavy chain of the immunoglobulin.

Of course, experiments of this kind were immediately done with the polyoma virus and SV40, and the results are similar to those obtained with the Rous virus: the immunoprecipitate formed by the anti-T serum with extracts of transformed or lytically infected cells also contains a protein-phosphorylating function. The difference, however, is that the immunoglobulin is not now phosphorylated, but the middle T is. The reasons for the different receptor specificities are not known. Undoubtedly these are very interesting findings; if we want to take these results at face value, we would say that the protein kinases specified by the different viruses are subjected in the cell to some types of restraints, probably different for the various viruses.

On the basis of these facts the pleiotropism of transformation can be explained better than it could at the beginning, although hypothetically. Probably there are at least three different reasons for the pleiotropism. One is that certain groups of cellular changes come from a common origin. Another factor is that in some cases there are several transforming proteins and each one may produce different types of alterations in the cells. The third factor is the virus-specified protein kinase. It is difficult to make predictions at this time, but this may turn out to be the most important factor in pleiotropism. In fact, by phosphorylating a number of cellular proteins, a protein kinase may affect many cellular functions. It is also conceivable that other viruses might specify some other kind of protein-modifying enzyme. Even with kinases, there may be a multitude of different kinds with different specificities.

7

The participation of such modifying enzymes in the phenotype of transformation might explain why transformation is very strongly influenced by the state of the cells and by the spectrum of cellular genes that are expressed in the cells. Of course, the reason is that the proteins modified by the enzyme are cellular.

This is the state of our knowledge of viral transformation today. As you see, there are some uncertainties--and it would be terrible if there weren't any--but of course the situation is very exciting because lots of interesting possibilities now loom on the horizon. What I want to do now is to try to look into the possible implications of all these findings for the more general problem of transformation by chemical or physical agents and for cancer.

It seems clear that mutations in the genetic material of somatic cells are a very important component in the mechanism of cancer. This conclusion is supported by a very large body of evidence. The question that I wish to discuss is whether there is any relation between the mechanism of cancer and that of viral transformation which I have analyzed. At first sight, the two processes seem to be basically different because viral transformation is due to the expression of genes imported by the virus, whereas chemical transformation-- and possibly also spontaneous transformation and cancer--is caused by changes in genes resident in the cells. However, this difference tends to disappear if we consider the origin of the viral transforming genes. The available evidence strongly suggests that the viral genes are of cellular origin but have undergone a special evolution for the virus' sake. Therefore, viral transforming genes must have counterparts in the uninfected cells.

This clearly is the case for the sarc gene of the Rous virus, which is largely homologous in its base sequence to a cellular gene. And the same is probably true for the transforming genes of other viruses, although they are certainly different from the sarc gene. So we must then consider that these normal cellular genes corresponding to the viral transforming genes may well generate cancer under certain conditions--namely, they are potential cancer genes.

Pursuing this line of thought, we can accept temporarily, for the sake of discussion, that chemical transformation and cancer are the results of the activation of such cellular cancer genes. Obviously, if a cell has a gene with such a potential, the gene must remain silent as long as the cells are normal. Therefore, there may be some mechanism of control, because in a cell homozygous for this particular gene the activation of only one copy might cause transformation. In fact the function of the gene would be dominant in the cell. So in order to minimize the occurrence of cancer, the gene must be negatively controlled, for instance by some kind of repressor specified by a pair of regulatory genes. If so, then cancer would require at least two mutations, one in each for the two regulatory genes.

Now, in fact, observations on the age-dependence of human cancers do suggest that cancer induction requires more than one event. This is deduced, for instance, from the slopes of the curves relating cancer incidence to age of exposure to an agent. At least two mutations are required for certain childhood neoplasias such as the retinoblastoma, and most adult cancers seem to require several.

The model that the activation of cellular genes normally repressed is responsible for chemically-induced or spontaneous cancers is supported by experiments with cell hybridization. I think the evidence is now really good that the hybrid cells formed by fusing a cancer cell and a normal cell are not neoplastic, provided they contain all the chromosomes of both parental cells: hence the cancer genes of the transformed cells seem to be repressed by products of genes of the normal cells. Therefore, it seems that virus-induced transformation, although a special case, shares some basic characteristics with spontaneous or chemically-induced cancer. In both cases, we can attribute the cellular changes to the activation of a cancer gene which perhaps specifies an enzymatic protein with multiple effects. The cancer gene is normally repressed in the cells but is active in the virus. Perhaps when this gene was incorporated into the virus, it was isolated from its controlling element or later underwent suitable mutations.

But why should an unruly protein kinase or another modifying enzyme cause cancer? The virus, I think, offers an explanation. Evidently, the presence of a transforming gene in the minimal genome of a very small virus, which goes through all these tricks to make three proteins out of one piece of DNA, must mean that the gene is essential for the reproduction of the virus. From this we can generalize that any cancer gene promotes replication of the cellular DNA and activates the related enzymatic machinery. In fact, that is what we found quite a long time ago for the transforming genes of polyoma virus and SV40. A protein-modifying enzyme is especially suitable for this function because it can change the behavior of, for instance, chromatin proteins, which directly control replication and expression of genes. It can also change the behavior of proteins that control replication indirectly--for instance, proteins of the plasma membrane or of the cytoskeleton.

The multiplicity of transforming viruses, each carrying a different trans-forming gene, suggests that there is a corresponding multiplicity of potential cancer genes in cells. A possible reason for this multiplicity is that different genes are used at different stages of differentiation for regulating normal cellular growth and only the genes normally active at a certain stage of differentiation may be able to become cancer genes. If so, the genetic and molecular mechanism of the different cancers would be different. This seems to be a rather general inference, because different types of cancer, whether chemically or virally induced, originate from cells in different states of differentiation.

As we have seen already, viral transformation emphasizes the importance of the cellular genetic background. The same probably applies to spontaneous or chemical cancers; especially in long-lived animals, as humans, the cells must have a whole series of genes blocking the consequences of the activation of a cancer gene and the number of blocking genes will be different for various cancer genes. A glimpse at this backup system is gained by observing the stepwise evolution of a spontaneous human cancer, which perhaps reveals the progressive dismantling of the defenses provided by the viral genes, one by one. Hence we can understand the variable and usually high number of mutations required for the expression of human cancers.

In conclusion, on this 50th Anniversary, we can contemplate with some satisfaction the progress made in the field of cell transformation and cancer. As in many other fields of biology, the experiments were initiated in these laboratories and then spread to many other places, forming a community of the plasma membrane or of the cytoskeleton.

Selected Readings

Colletee, M. S. and R. L. Erikson. Protein kinase activity associated with the avian sarcoma virus src gene product. Proceedings of the National Academy of Sciences, Washington, **75**:2021-2024 (1978).

Oppermann, H., A. D. Levenson, H. E. Varmus, L. Levintow and J. M. Bishop. Uninfected vertebrate cells contain a protein that is closely related to the product of the avian sarcoma virus transforming gene (src). Proceedings of the National Academy of Sciences, Washington, **76**:1804-1808 (1979).

Sambrook, J., H. Westphal, P. R. Srinivasan and R. Dulbecco. The integrated state of viral DNA in SV40-transformed cells. Proceedings of the National Academy of Sciences, Washington, **60**:1288 (1968).

Stanbridge, E. J. Suppression of malignancy in human cells. <u>Nature</u> **260**:17-23
  (1976).
Tijan, R. and A. Robbins. Enzymatic activities associated with a purified
  simian virus 40T antigen-related protein. <u>Proceedings</u> <u>of</u> <u>the</u> <u>National</u>
  <u>Academy</u> <u>of</u> Sciences, Washington, **76**:610-614 (1979).
Tooze, J. The <u>Molecular</u> <u>Biology</u> <u>of</u> <u>Tumor</u> <u>Viruses.</u> Cold Spring Harbor
  Laboratory: Cold Spring Harbor, New York (1979).

RNA VIRUSES AND CANCER

Howard Temin, Nobel Laureate

McArdle Laboratory for Cancer Research
University of Wisconsin, Madison

Twenty years ago when I was a graduate student at Caltech, I started working with the Rous sarcoma virus in Renato Dulbecco's laboratory. Rous sarcoma virus is the prototype of a family of viruses that are now called retroviruses, because they do things a little backwards. One feature some of them have is the ability to transform fibroblasts into tumor cells. This transformation phenomenon is just one remarkable feature of Rous sarcoma virus.

The other remarkable feature of Rous sarcoma virus and of retroviruses in general is shown by nucleic acid hybridization experiments. These are experiments in which labeled viral RNA, for example, from a virus called Rous-associated virus-0, is annealed to the DNA of, for example, normal chicken cells, in this case the kind of chicken you would otherwise eat. There is great difficulty in this field in defining normal, so I am defining normal as something you would eat.

When these experiments were done in the late 1960's and early 1970's, it was found that there were nucleotide sequences related to the virus in the uninfected chicken cell DNA. However, none of this Rous-associated virus-0 DNA could be found in duck cells. Thus all cells did not have this virus-related DNA, but only cells related to chickens had it. In fact, Peter Vogt showed that normal "uninfected" cells from some special chickens were actually producing this virus. In this talk, I want to bring you up to date on some of the processes that may be involved in the evolution of these viruses and in their replication. In order to do this I will first give you a little background about these viruses.

First, I need to indicate that most viruses are not like Rous sarcoma virus. Rous sarcoma virus is not even a typical retrovirus, that is, a virus that has RNA and replicates through a DNA intermediate. Rous sarcoma virus is really one of a very special kind of laboratory artifact which I call the strongly transforming or rapidly oncogenic viruses. These are viruses that are very useful to study. They persist in laboratories because they are so interesting, and thus they are evolutionarily very successful. Some, like the Rous sarcoma virus, are not defective. All the others, for example, the avian myeloblastosis virus, myelocytomatosis 29 virus, are defective in their replication. They need a helper virus to replicate. Of the retroviruses that are more prominent in nature (nature being whatever is outside the laboratory, because for a chicken there is really not nature outside of a farm or a poultry processing plant), the oncogenic virus one normally finds is what I call weakly transforming or slowly oncogenic viruses. These may require months to cause cancer. Examples are lymphoid leukosis virus of chickens, murine leukemia virus of mice, and feline leukemia virus of cats. Perhaps the largest number of retroviruses are those that are not oncogenic. These like the weakly oncogenic viruses are not defective. Examples are Rous-associated virus-0 whose nucleotide sequences are found in normal chickens, spleen necrosis virus, a cytopathic virus, most of the endogenous murine leukemia viruses, and other endogenous viruses.

As with polyoma virus, a great deal of mapping has been done with retroviruses, and it is possible to draw a rough map of the gene order for these viruses. There are many special features in such a map which do not need to concern us in this discussion. In nondefective retroviruses there are three major genes--a gene for the internal proteins (gag), a gene for the DNA polymerase (pol), and a gene for the envelope glycoprotein (env). In some cases the products of these genes are then cleaved so there is more than one protein per gene. The strongly transforming, nondefective Rous sarcoma virus has an additional gene, src, which controls the formation of a protein kinase. The majority of strongly transforming RNA tumor viruses, however, are very different from Rous sarcoma virus in that they are defective; they cannot replicate by themselves, but are found with another virus that supplies several necessary gene products. One usually finds quite a scrambled genome in these defective viruses. It usually has a 5'-end related to the nondefective viruses followed by new nucleotide sequences and at the envelope region also altered nucleotide sequences.

With this background, then, I want to present a series of hypotheses, which are called protovirus hypotheses, concerning these viruses and their relationship to the cell genome. (1) The strongly transforming retroviruses evolved from weakly transforming retroviruses and cellular DNA. This hypothesis has been established for most of the strongly oncogenic viruses, for example, Rous sarcoma virus comes from lymphoid leukosis viruses and chicken DNA; Kirsten murine sarcoma virus comes from murine leukemia virus and rat DNA; Moloney murine sarcoma virus comes from murine leukemia virus and mouse DNA; and spleen focus-forming virus comes from murine leukemia virus and mouse DNA.

This hypothesis is supported by experiments carried out in San Francisco where a specific cDNA was prepared to the src region of Rous sarcoma virus. To do this, cDNA was made complementary to Rous sarcoma virus RNA which means there was cDNA complementary to the gag, pol, env, and src genes. Then, this cDNA was hybridized to the RNA from a lymphoid leukosis virus that contains all these genes except src. The DNA that was not hybridized was separated from the rest of the DNA that was hybridized. A specific probe was found that only annealed to a virus containing the transforming gene, src.

When such DNA is hybridized to the DNA of Rous sarcoma virus-infected cells, there is a high melting temperature indicating a good match, as one would expect. Hybridizing to the DNA of other cells, for example, chicken cells, the src DNA is found to be related to chicken DNA, but not identical. The amount of hybridization is not as high, and the melting temperature is decreased, indicating that there is some mismatching. When DNA comes from birds other than chickens, the amount of hybridization and the melting temperature further decrease, showing greater mismatching. It appears from these kinds of data gotten with many kinds of viruses that at some time there was a recombinational event, plus mutational events, between a weakly oncogenic virus and some cellular genetic material, which gave rise to the strongly transforming virus. These strongly transforming viruses normally would not persist in nature because they kill their hosts and there is no way for them to spread. However, scientists have passed these viruses and so they persist.

We can suggest, and there is evidence to back up the hypothesis, that (2) the weakly transforming viruses--which, as I indicated, do not contain specific transforming genes--cause neoplasia by forming strongly transforming genes. This formation would happen in the few months that it takes for one of these viruses to cause cancer. The best example is from Wally Rowe's laboratory: where the endogenous AKR murine leukemia virus has been shown to change to a mink-cell focus-forming virus and transforming virus. Similarly the endogenous C57Bl murine leukemia virus changes to a transforming radiation leukemia virus, a thymotrophic murine leukemia virus which appears to be strongly transforming. Thus, in the case of viruses that are strongly transforming, the evolution to oncogenicity has happened in the past; while, in the case of the weakly transforming viruses, this evolution occurs in a short time span and can happen very reproducibly by stochastic processes in a single animal.

Further evidence, of the same general nature that I have mentioned for the origin of strongly oncogenic viruses, especially the evidence of the nucleic acid hybridization being greatest and the nucleotide sequences being most

12

homologous in the species of origin, while in related species, those nucleotide sequences are found in lesser amounts and in less related forms has led to the hypothesis that the weakly transforming retroviruses evolved originally from cellular DNA. Examples would be the avian leukosis viruses, which I called previously lymphoid leukosis viruses, evolved from chicken DNA, and the murine leukemia viruses originated from endogenous sequences in mouse DNA.

This hypothesis is supported by a good deal of nucleic acid hybridization evidence. It has even been possible, by a combination of nucleic acid hybridization and determining the infectivity of different viral DNAs, to find stages in the escape of a virus from the cell genome, stages in which cellular DNA sequences are getting ready to escape and become an independent genetic system.

In replication, the virus forms a provirus. Viral infection produces DNA with the sequences of the virus and these sequences are transcribed, giving rise to a large amount of virus. This DNA is infectious: if one isolates the cell DNA and adds it to an uninfected cell, virus is produced. There are certain normal chicken cells that contain by nucleic acid hybridization similar nucleotide sequences to those of infected cells. Cells exist that produce very small amounts of virus. The DNA from these cells is not infectious. This result has been interpreted as indicating that these cells have some cis-acting control element that is attached to the DNA of the virus.

Other chicken cells also appear to have similar nucleotide sequences, but these sequences do not make virus and the DNA is not infectious. In addition, normal cells contain many viral-like products: reverse transcriptases have been isolated from chicken and goose cells; viral glycoproteins are distributed in certain organs; and internal proteins of the virus are present in other cells.

It appears that these genes are present and expressed in normal cells and that through time some of them have gotten in a package and then out to what we call a virus.

More recent work (Cohen and Varmus, Hughes et al., Astein et al.) indicates that some of the endogenous viruses, including the RAV-0 in chicken cells, are a result of recent germline infections. Thus, this particular evidence cannot be used in support of virus evolution originally from cell DNA. However, I still think that is an attractive hypothesis for the original appearance of the different genera of retroviruses.

We can also suggest that processes similar to these involved in viral oncogenesis and evolution are involved in non-viral cancers. The evidence for this hypothesis is as yet weak. There may be formation of similar types of genes, as Dr. Dulbecco has also suggested, and the formation of these genes may be related to processes like those involved in the evolution of the retroviruses that I mentioned earlier.

The most remarkable results that support this hypothesis have been published in 1978 from the laboratories of Rasheed, Gardner and Huebner, and Rapp and Todaro. They passaged a nontransforming endogenous virus with normal cells. The normal cells became spontaneously transformed, and the virus varied genetically. When they took this activated virus, which was apparently nontransforming and nononcogenic, and infected transformed cells (cells that were transformed by chemicals or spontaneous tumor cells), they recovered a transforming retrovirus. In some cases, these viruses were also oncogenic.

This experiment indicates that these transformed cells contained genes that were able to undergo some kind of phenotypic or genetic interaction with the original virus to give rise to transforming retroviruses. So it seems that there may be some continuity in processes involved in virus evolution, virus carcinogenesis, and nonviral carcinogenesis. Dr. Dulbecco has also indicated that this continuity may extend to the proteins controlled by these different genes.

So, retroviruses are very different from other viruses in having these remarkable relationships with the cell genome. These relationships are the result of the fact that they can integrate their viral information with cellular DNA, synthesize RNA from the cellular viral DNA, and in turn make more viral DNA. This ability apparently gives them a remarkable genetic plasticity.

Recently, much has been learned about the integration part of this process--that is, how foreign extrachromosomal viral DNA gets integrated with cellular DNA. I would like now to describe some of the recent results about the mechanism that retroviruses use to integrate their DNA. In particular, I want to address these questions (questions which were asked many years ago about temperate bacteriophage): Is the viral DNA integrated at a unique site in cellular DNA or are there multiple sites? Is there a unique site for integration in viral DNA or multiple sites for integration? And finally, does the site of viral integration affect viral functions or do viral DNAs integrated at different sites have the same properties?

I am going to approach these questions with a virus named spleen necrosis virus, a member of the same genus of avian retroviruses as Rous sarcoma virus. Other people have gotten roughly similar answers using Rous sarcoma virus. Now spleen necrosis virus does not transform cells; it causes a lytic cellular infection and an acute infection with necrosis in animals. In a viral growth curve (Figure 1), following a latent period, virus begins to be produced and then

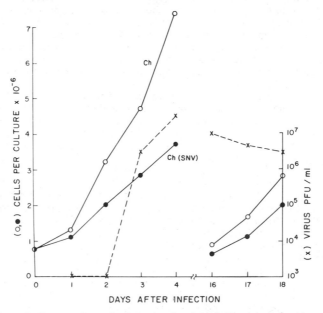

FIGURE 1 — *Cell growth and virus production following infection of chicken embryo fibroblasts by spleen necrosis virus. Ch, mock infected cells; CH(SNV), spleen necrosis virus-infected cells.*

is produced continually. Growth of normal cells compared to growth of infected cells shows that some of the infected cells are killed at this point, early after infection. This phase is called the acute phase of infection. After 10 days, the normal cells and the infected cells multiply at the same rate. The infected cells look like the normal cells, but they continue to produce virus. This phase is called the chronic phase of infection.

14

The experiments have involved two different ways of looking at viral DNA. One is biological and the other is nucleic acid hybridization. I want to first indicate the steps we used in getting "Southern blots." We infect a large population of cells with virus so we are seeing the result of a large number of separate infections. We extract DNA and then the DNA is digested with restriction enzymes. The DNA fragments are separated in agarose gel electrophoresis on the basis of size. At this point two things can be done: Either we can take these DNAs and assay them for their biological infectivity, because as I told you, the DNA of these viruses is infectious; or the DNA can be denatured, transferred by the method of Southern onto nitrocellulose filters, hybridized with virus-specific probes (either iodine-labeled viral RNA or ^{32}P-labeled cDNA), and then visualized by autoradiography, giving the so-called "Southern blots" shown in Figure 2.

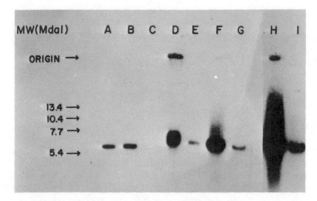

FIGURE 2 — *Forms of spleen necrosis virus in infected chicken cells early and late after infection.*

First we can look at the two forms of viral DNA in an infected cell. The major form is a linear, unintegrated molecule of about 6,000,000 daltons, which is a copy of the viral RNA into a double-stranded DNA. The other form of viral DNA found in cells, if there is no digestion, is a very large viral DNA. This is viral DNA integrated with cellular DNA and so it is in molecules larger than 30,000,000 daltons and does not enter the gels.

We found that the common restriction enzyme Eco RI did not cut this virus DNA. Therefore, we could cleave the infected cell DNA with Eco RI and see how viral DNA was distributed in terms of the Eco RI sites in cellular DNA. After exposure to the enzyme, we found that viral DNA is found in a great many molecular size classes. In Figure 3, viral DNA is shown by zig-zag lines integrated with cell DNA shown by the straight lines. Eco RI sites are indicated by arrows. If the virus is integrated at a unique site in the cellular DNA, it will be in all the cells at the same distance from Eco RI sites, so after digestion and assay we should find a single band of hybridization. However, if the virus is integrated at multiple sites, there should be a difference in where the Eco RI sites would be in relation to the viral DNA and we should find more than one hybridization band. As can be seen in Figure 2, what we found was compatible with the latter model, where there are actually many more than three different sites. The viral DNA is isolated from these cells early after infection and is at multiple sites. Before digestion all the infectivity is in very large molecules; after digestion with Eco RI the infectivity is spread over a variety of size classes of molecules, again indicating that infectious viral DNA, and, therefore, the whole viral genome, is integrated at multiple sites soon after infection.

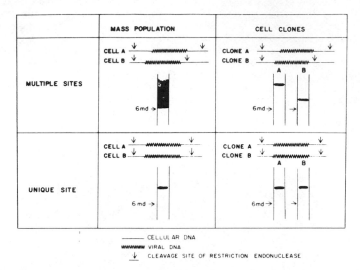

FIGURE 3 — *Multiple or unique sites of integration of viral DNA in cell DNA: cell sites.*

However, we found a different answer looking at cells late in infection, after the cytopathic effect and cell killing has disappeared. Looking at DNA from chronically infected cells (Figure 2), one can see that there has been a great decrease in the amount of integrated DNA. In fact, there is a 7-fold decrease in the amount of integrated DNA and an almost 100-fold decrease in the amount of unintegrated DNA late in infection. This decrease is correlated with cell death and may actually cause it.

Figure 4 illustrates our results following a longer exposure, comparing unintegrated DNA, uninfected chicken cell DNA, and nucleic acid hybridization with the Eco RI digest of DNA from chronically infected cells. Again we see a smear indicating that viral DNA is also in these cells integrated in multiple sites. However, when this gel is sectioned and infectivity measured, one finds that the infectivity is present in only one of the size classes of DNA, which is quite a difference from acutely infected cells. In the acutely infected cells, viral DNA is integrated in multiple sites and all of the sizes of DNA are infective; in the chronically infected cells, however, viral DNA is still present at several sites in the cell, but there is less of it, and the infectious DNA is only found at a single site.

This raises, then, the question of what is the difference between viral DNA at these infectious and noninfectious sites? To answer that question we required a restriction enzyme map of viral DNA. Figure 5 shows that we can take viral DNA and cut it with specific restriction enzymes. We also found that the unintegrated viral DNA is colinear with viral RNA. With these enzymes and such a map we are then able to ask about the integration sites in the virus. We have already shown that there are multiple integration sites in the cell; now we want to ask about the integration sites in the virus. For this there are three known possibilities. In Figure 6 we show an example of unintegrated viral DNA that is cut twice by a restriction enzyme, giving rise to three fragments with their size in megadaltons. Lambda phage integration is a result of circularization of the genome and integration somewhere in the center of the linear

16

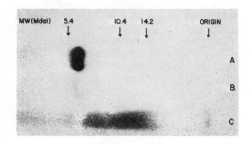

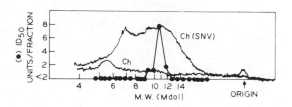

FIGURE 4 — *Infectivity and nucleic acid hybridization of DNA from chicken cells chronically infected with spleen necrosis virus.*

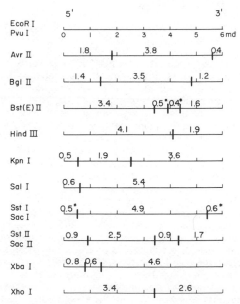

FIGURE 5 — *Map of restriction enzyme cleavage sites of linear unintegrated spleen necrosis virus DNA.*

17

map, giving rise to a permuted map. If the integrated DNA is cut by this restriction enzyme, the terminal junction fragments will be spread through the gels, because these fragments, as a result of the multiple sites of integration, are different in size. The unique fragment found is a fusion fragment of the 3'-end and the 5'-end, a new fragment that is not seen in unintegrated DNA.

Now the papova virus, of the type that Dr. Dulbecco talked about, starts with a circular genome. Apparently integration can take place in any place in the virus or in the cell, at least for SV40. Therefore, there will be many different possibilities, three of which I have put down in Figure 6, whether integration takes place in the center region, which is largest, or in the 5'- or 3'-end. What would be seen after digestion would be the internal fragment of unintegrated DNA and the fusion fragment of the two ends.

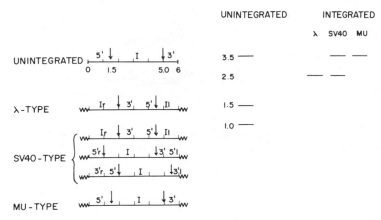

FIGURE 6 — *Possible modes of integration of viral DNA in cell DNA: viral sites.*

Finally, there is the type of integration, seen with bacteriophage mu, where the viral integration does not permute the map and the termini of the virus seem to be responsible for integration. This does not say that there cannot be a circular intermediate--it says nothing about the mechanism of integration, only that the map is still colinear to linear viral DNA.

This experiment has been carried out with a variety of enzymes. We will look at this combination of Sac I and Kpn I (Figure 7). Starting at the 5'-end of the unintegrated DNA, we first find a fragment of 500,000 daltons, then another fragment of 1.9 million daltons, then a fragment of 2.9 million daltons, and then again a terminal fragment of 0.6 megadaltons. When this same enzyme combination is used to digest integrated DNA, we only see the internal fragments, we do not see any fusion fragments. We can quantitate the ratio of these fragments and compare the amount expected if the integration were colinear. We find that within the limits of our measurements, which have an accuracy of about 10%, all of the integrations are colinear. This experiment and others like it, indicate quite strongly that the integration of the virus is at a unique site in the viral DNA and that this site is at or very near the termini of the virus. There actually appears to be a terminal repetition of 640 base pairs at the end of the virus that may be important for this integration. In terms of the bacteriophage results, this would be an integration like that of phage mu.

Knowing that we can transfer the map of unintegrated viral DNA to that of integrated viral DNA, we can answer the question, at least to a certain level

18

of resolution, of whether or not there are gross abnormalities in the non-infectious integrated viral DNAs. As you will remember, there were DNAs in the chronically infected cells, where the Eco RI cut DNA was larger, that were noninfectious. One might imagine they had some kind of insertion causing this lack of infectivity. Then there were other DNAs from these cells that were smaller and were also noninfectious. One could imagine deletions in these. These would be fairly large insertions or deletions, so we carried out experiments to ask the question: Are the maps different for the noninfectious integrated viral DNAs?

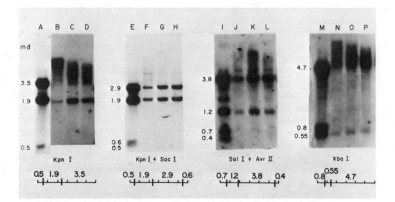

FIGURE 7 — *Restriction enzyme digestion of noninfectious and infectious fractions of Eco RI-digested DNA from chicken cells chronically infected with spleen necrosis virus.*

Figure 7 shows what we have found. There are actually four different experiments. Southern blots are shown. In each case there are four columns: the first one is the pattern of digestion of unintegrated viral DNA with the enzyme indicated below, Kpn, Kpn I plus Sac I, Sal I plus Abr II, and Xba I, giving rise to all of the fragments that were indicated on the earlier maps and below the autoradiograms. Then we took the DNA from chronically infected cells, digested it with Eco RI, separated the fractions on the basis of size into noninfectious fractions, smaller than 9 million daltons, infectious fractions, and noninfectious fractions, larger than 10 million daltons. We digested these DNAs with these enzymes which cut the viral DNA, and asked whether or not the internal fragments, which are all we would expect to see, were different. We might expect deletions in the smaller noninfectious molecules and insertions in the larger noninfectious molecules. We chose enzymes that cut at different spots all through the viral genome, so we looked for these deletions and insertions all through the genome. We see in the case of Kpn I that there is one internal fragment in all the samples. (Digestion was not quite complete in lane B.) Where the junction fragment with the cell is large, we see it is composed of a 3-1/2 million viral fragment plus cellular sequences. If there had been a deletion in the viral fragment we would have expected it to move faster. Or, if there had been insertions, we would have expected the junction fragment to be larger than 3-1/2 million daltons. Sac I cuts just 600,000 from the 3'-end, so one can look at the DNA which was in the junction fragment after Kpn I digestion. Looking at that DNA in the 2.9 million dalton band, we again see that there was no change; there were no deletions or insertions.

We have sampled a good fraction of the genome. Looking at many other

19

gels, we see no evidence for deletions or changes in the internal fragments. Finally, in the large terminal fragment of 4.7 million after Xba I digestion, we see a sharp cut-off right at 4.7 million, and we do not see as a result of a deletion that the band is migrating earlier or an insertion that it is running slower. So, in these experiments with a variety of enzymes, sampling throughout the genome, and in the noninfectious fragments, we were not able to find any gross abnormalities. Certainly we found no indication that there were large insertions or deletions.

We now can answer a few of the questions that I earlier posed about the integration process. Viral DNA is integrated at multiple sites in cellular DNA as defined in terms of Eco RI cleavage sites. Viral DNA is integrated at a unique terminal site in viral DNA. We have not shown that it is exactly at the terminus because we do not have enzymes that cut that close, but it is likely that it is. Finally, the viral DNA integration at different sites in cellular DNA may lead to noninfectious or infectious viral DNA. To explain this result, we hypothesize an influence of the neighboring sequences on the expression of the virus. Further, the integration at some sites may directly cause cell death.

Thus, these experiments indicate that retroviruses have specific methods for integration, very specific so far as the virus is concerned, but not quite as specific so far as the cell is concerned. However, the virus is apparently not completely autonomous once it is integrated, but is dependent for its expression on cellular nucleotide sequences. These types of experiments reveal molecular mechanisms for the events discussed in the protovirus hypotheses that I discussed with you earlier, that oncogenic viruses evolve from cells, and that similar processes, related to viral integration and replication, are involved in nonviral carcinogenesis and perhaps in differentiation. Of course, these hypotheses require much more experimentation to determine if they are correct or what part is correct.

Selected Readings

Bishop, J. M. Retroviruses. Annual Review of Biochemistry 47:35-88 (1978).
Keshet, E., J. O'Rear and H. M. Temin. DNA of noninfectious and infectious integrated spleen necrosis virus (SNV) is colinear with unintegrated SNV DNA and not grossly abnormal. Cell 16:57-61 (1979).
Keshet, E. and H. M. Temin. Sites of integration of reticuloendotheliosis virus DNA in chicken DNA. Proceedings of the National Academy of Sciences, Washington, 75:3372-3376 (1978).
Keshet, E. and H. M. Temin. Cell killing by spleen necrosis virus is correlated with a transient accumulation of spleen necrosis virus DNA. Journal of Virology 31:376-388 (1979).

ENVIRONMENTAL CHEMICALS CAUSING
CANCER AND MUTATIONS

Bruce Ames

*Department of Biochemistry
University of California, Berkeley*

Almost every week we open our newspaper, and there's another carcinogen of the week. We read about kepone fouling up Chesapeake Bay, and then it turns out to be a potent carcinogen. Or PCB turns up in Antarctic penguins. So I'd like to discuss how one detects carcinogens, and the relation between carcinogens and mutagens, and try to put this whole problem in perspective.

How do we detect chemicals that cause cancer? The traditional way has been to do it in people. Men started smoking cigarettes back around the turn of the century, and lung cancer in males followed along about 20 or 25 years later (see Figure 1). Women started smoking heavily around the time of the Second World War, and smoking in young women now is almost up to smoking in men. And now lung cancer in women all over the world is shooting up.

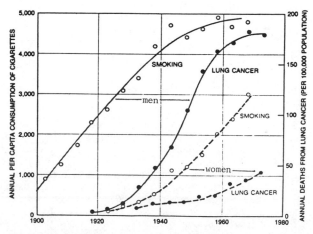

FIGURE 1 — *Relation between cigarette smoking and lung cancer. Cigarette smoking and lung cancer are unmistakably related, but the nature of the relationship remained obscure because of the long latent period between the increase in cigarette consumption and the increase in the incidence of lung cancer. The data are for England and Wales. In men (solid line) smoking began to increase at the beginning of the 20th Century, but the corresponding trend in deaths from lung cancer did not begin until after 1920. In women (broken line) smoking began later, and lung cancers are only now appearing. [From J. Cairns, "The Cancer Problem," *Scientific American* (November 1975), page 72. Reprinted by permission. Copyright c 1975 by Scientific American, Inc. All rights reserved.]*

The problem with environmental carcinogens is really that you don't see much for 20 to 25 years. After factory workers got exposed to beta-naphthylamine, 25-30 years later bladder cancer starting coming in. It was the same with the workers exposed to benzidine, or the radium dial workers, or after

21

Hiroshima--most of the tumors didn't start coming for 20 years. So by the time you discover something in people, it's already mostly too late.

Another problem is that it's fiendishly difficult to do human epidemiology. Twenty-five percent of us are getting cancer; if I get cancer tomorrow, is that due to an X-ray I got as a child, or some food additive, or pesticides in my body fat? It's really very hard to establish cause and effect, and clearly we're not going to keep up with the 50,000 chemicals in commerce by testing them in people.

Chemicals are made on a large scale in the modern world. I'll just use two chemicals as an illustration. Both vinyl chloride and ethylene dichloride are major chemicals (see Figure 2), and they are both mutagens and carcinogens.

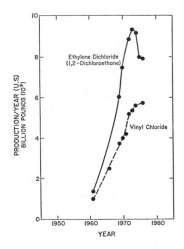

FIGURE 2 — *Production of two mutagens and carcinogens with widespread human exposure: ethylene dichloride and vinyl chloride (production data from "Top-50 Chemicals" issues of* Chemical and Engineering News*). Approximately 100 billion pounds (5×10^{10} kilograms) of ethylene dichloride and over 50 billion pounds of vinyl chloride have been produced since 1960. Ethylene dichloride is a volatile liquid that is the precursor of vinyl chloride and is also used extensively as a fumigant, solvent, gasoline additive (200 million pounds per year), and metal degreaser. Ethylene dichloride was first shown to be a mutagen in* Drosophila *in 1960, and later in barley and* Salmonella *but these data have been ignored. The first adequate cancer test in animals has just been completed by NCI (September 1978) and shows ethylene dichloride to be carcinogenic in both sexes of rats and mice. Vinyl chloride gas is used to make polyvinyl chloride (PVC; vinyl) plastic. It was shown to be a carcinogen in rats and in people in the mid-1970's, and a mutagen in* Salmonella *and other systems shortly afterward. Vinyl chloride production results in the dumping of enormous quantities of a waste product, EDC-tar, that is a complex mixture of chlorinated hydrocarbons; this is mutagenic in* Salmonella *and has been detected as an ocean pollutant.*

Vinyl chloride was being made at a level of 4 billion pounds a year in the U.S. when someone tested it in animals and also found that it is a carcinogen in people. We didn't ban vinyl chloride; we just started treating it with respect. It was so cheap, 9 cents a pound, that it was used in millions of spray cans, and women were spraying their hair in bathrooms and they (and factory workers) were breathing in 300 parts per million of vinyl chloride. In addition, there was lots of monomer dissolved in the polymer and we were walking around on vinyl floors, and using PVC pipe and eating food packaged in vinyl. Formerly, the monomer was getting into food and into people, but now we make sure that the monomer is removed from the polymer. We started treating vinyl chloride with respect, but we didn't ban it.

In the same way, we are not going to ban many of the carcinogens, but we want to know they are carcinogens. Ethylene dichloride is the precursor of vinyl chloride and is one of the major organic chemicals in the U.S. It's currently being made at the level of 10 billion pounds a year. It was first shown to be a mutagen in Drosophila, and a fairly potent mutagen, back in 1960. Then it was shown to be a mutagen in barley, and in our Salmonella tests, and now the first cancer test has just come out of the National Cancer Institute, showing that it's a potent carcinogen in male rats, female rats, male mice, and female mice. This is 100 billion pounds later.

It's clear that there must be some middle ground between total neglect and banning all carcinogens. Clearly we do not want to repeat this--making 10 billion pounds a year of a chemical and not knowing that it's a carcinogen, when there's a tremendous amount of human exposure.

It should be remembered that the large-scale production of these man-made chemicals started in the mid-fifties and went up fast, so whatever cancer they are going to cause will hit us in the 1980's. I think you could make a pretty good argument that very little of the cancer today is in fact due to the modern chemical world.

Some people say, "Well, are all these chemicals getting into us?" It's clear that, at least for the first generation of pesticides, they did. They were mostly chlorinated chemicals that are very nonpolar and are bioconcentrated. The amounts in people are small, but because they bioconcentrate they are in everybody's body fat. Some are fairly weak carcinogens like DDT, and others are more potent like dieldrin. The effect of all this is that we don't know; we'll just have to see.

Then there's the question, "Is there a safe dose? Maybe all this is a big scare, and we don't have to worry about these low doses." Nobody really knows the answer, because if you do an experiment with 30 rats, you can't obtain a decent dose-response curve. That is, nobody knows what is the effect of one-one thousandth of the amount of chemical that's giving the rats and mice cancer, because of the problem with statistics and small numbers of animals. Nobody can tell you what the shape of that dose-response curve is. My own feelings are that it's probable some chemicals would give linear curves, and some would look like hockey sticks, but right now we just can't answer that. I think it's prudent, though, to bear in mind the point that Richard Peto and others made, that 25% of us are getting cancer, and that we're way up some dose-response curve already. I'll discuss why it looks as if many of these chemicals are working in the same way as mutagens.

Why not test chemicals in animals? It does seem a little strange to make 10 billion pounds a year of something and not do an animal cancer test. But it costs over $200,000 to do an animal cancer test and it takes two or three years. In addition, the drug companies go through 100 chemicals before they finally get to one they like. They've spent five years and put in millions of dollars, and by the time they finally get that perfect tranquilizer they really are not overjoyed to find out that it is a carcinogen. They need some way of looking at the chemicals along the way. And the same thing goes for chemical companies. An animal cancer test just takes too long and is too expensive.

We're surrounded by mixtures of carcinogens: There are a large number of carcinogens in the natural world, there are mold toxins, carcinogens in plants, and you're pouring carcinogens out of your car exhaust. How do we identify all these mixtures? Again, animal cancer tests really can't be used as a bioassay.

There are 50,000 or so chemicals in commerce and 1,000 new ones coming out each year, and there are just not enough veterinary pathologists to even keep up with DuPont. There are so many chemicals and mixtures in the environment that we need something more than just a cancer test in animals.

There is one last problem with animal cancer tests. If you have a hundred million people exposed to a carcinogen, you'd like to be able to pick up a chemical that caused a 1% increase in cancer. If there were a million extra cases of cancer, you'd want to know about it. But there's no rat or mouse experiment you can do that will pick up a 1% increase because of the problem of statistics and small numbers of animals. Nobody wants to do an experiment with more than 50 animals in a group. It's too expensive. So they fill the animals with as much of the chemical as they can, just below the dose that will kill them, and then they do the experiments. All cancer experiments in animals are done that way, trying to overcome the problem of small numbers. But even then, there are problems in sensitivity, and if it comes out as a carcinogen someone will say, "Well, you have to drink ten bottles of hair dye, or 500 soft drinks, or eat a pound of DDT."

What I'd like to talk about now are the new short-term tests that have been coming up that are still another way of looking at the problem and can be used in addition to human epidemiology and animal cancer.

About 15 years ago when I was at NIH--working in the Arthritis Institute, incidentally--I became interested in the problem of birth defects in people. Everybody who comes out of Caltech Biology comes out half a geneticist anyway, and I started wondering what would happen if the human gene pool suddenly started getting filled with mutations. If some of the food additives, or whatever, turned out to be mutagens like nitrosoguanidine, would we ever discover it? So we started thinking about the histidine mutants that Philip Hartman and I had worked with in bacteria. These are mutant bacteria Salmonella typhimurium with a defect in one of the genes for histidine biosynthesis. Now if we treat them with a mutagen, we can mutate some of the bacteria back to normal; they can then grow on a minimal medium and give rise to colonies, which we can count. We started looking at all the known mutagens and developed a test system. We incorporate the chemical into the agar, do a series of concentrations, and subtract the number of colonies on the control plate from the number of colonies on the treated plates, and practically always we get a linear dose response. Mutagenesis is linear for 95% of the chemicals. One of the strains we use happens to have a long repetitive sequence in its DNA: CGCGCGCG--a particular hot spot for frameshift mutagenesis. We stumbled on this after screening hundreds and hundreds of mutants for reversion by a particular mutagen.

Over the years we improved the sensitivity by eliminating DNA repair in the bacteria and peeling off the outside lipopolysaccharide, so the bacterium became more like an animal cell. As we did this, we kept on looking at different chemicals, and more and more I became interested in the idea of the relation between mutagens and carcinogens. If we just take chemicals off the shelf, very few chemicals are mutagens. They cannot interact with DNA, yet, many mutagens were known as carcinogens. In fact X-rays and mustard gas were known as mutagens in Drosophila before they were shown to be carcinogens. The first theory of cancer was that chemicals were causing cancer because they were causing somatic mutations.

Each cell has programmed into its DNA when to grow and when not to grow. Your liver grows to a certain size; you cut your skin, and it fills in and then stops. The first idea, from Boveri on, was that maybe these chemicals were damaging the genetic material and causing the cell to become mutant in some way and continue growing when it wasn't supposed to. That was a very attractive theory, particularly to a molecular biologist, so we started looking at many of the chemicals known to be carcinogens to see if they were mutagens. In fact, a very high percentage of them turned out to be mutagens.

We kept on improving the bacterial strains but there were many chemicals that weren't shown to be mutagens and yet were known as carcinogens. A very good example is benzpyrene, which is a chemical you get whenever you burn anything. You are pouring benzpyrene and its relatives out of your car exhaust when you drive, particularly if you drive a diesel car. Whenever you burn things, you make polycyclic hydrocarbons. Many years ago people showed that if you paint it on the skin of animals you get tumors, but benzpyrene wasn't a mutagen. Then a big advance in the field came when people like Boyland, Sims and Grover, and the Millers started learning that the true carcinogen was a reactive metabolic derivative. If you incubate benzpyrene with liver, it is converted to many different chemicals, and it was shown that in fact the true carcinogen is an epoxide of benzpyrene.

We had been interested in these active forms, and I had written to the Millers and was sent some active forms of aromatic amines that they had worked with, and from Sims and Grover I got some active forms of polycyclic hydrocarbons. We showed that these in fact were mutagens in our system, and I became more and more convinced about the relation between mutagens and carcinogens. But obviously, if one wanted to detect carcinogens by the bacterial test, some way of introducing mammalian metabolism was needed, because bacteria don't do this kind of metabolism. They will pick up a direct alkylating agent, but they won't respond to a precarcinogen like benzpyrene. But people who incubated benzpyrene or other chemicals with extracted liver showed that you could get active forms. What we did then was to grind up some liver and put it right on our petri plates. The liver is then a first approximation of a rat, and the bacteria provide a sensitive detection system for damaged DNA.

We were finding that more and more chemicals were showing up as mutagens when we did this, but we always had a few chemicals that weren't working as mutagens that really should have worked. For example, aflatoxin is a supercarcinogen that is made by a mold, Aspergillus flavus; it is one of the most potent carcinogens known. We were finding it was a mutagen, but it was very weak; similarly some of the nitrofurans were very good carcinogens and weren't showing up as mutagens. And then we made another improvement in the test system, which had to do with error-prone repair, but I won't go into that. Now we found that these chemicals worked extremely well. We can detect a nanogram of aflatoxin, and again one gets a linear dose-response curve. When we first started working with it, we were worried about handling all these nasty carcinogens. We were handling a few nanograms of aflatoxin, finally burying our waste in the desert somewhere, through the University Office of Health and Safety and all of that--until I finally calculated how much aflatoxin the Food and Drug Administration allowed in a jar of peanut butter, and it's 15 micrograms.

Joyce McCann and I then decided to see if we could validate the test system. How good was ground-up rat liver plus Salmonella at detecting carcinogens? One would like to look at the known human carcinogens, and in fact if you take a look at the list of organic chemicals that are known or suspected to cause cancer in people it's a fairly short list, not because many chemicals don't cause cancer in people but because of the difficulty of studying human epidemiology. Almost all of these carcinogens show up as mutagens in the test system: vinyl chloride, nitrobiphenyl, beta naphlamine, mustard gas, cigarette smoke condensate, bis-chloromethyl ether, aflatoxin, and others. We missed benzene and diethylstilbestrol.

But then we wanted to look at a much more detailed collection of chemicals. There are many chemicals known to cause cancer in rats and mice, so we looked at a large number of chemicals--300 altogether. Out of 175 carcinogens, we found that 90% were mutagens in the system; we missed 10%. About this 10%--some of these we can't detect because they are chemicals like diethylstilbestrol, which is a hormone analog, and I think steroids are probably working in a different way, not by interacting with DNA. Also, there are chemicals like griseofulvin, which is a carcinogen and an antibiotic drug. Griseofulvin is known to interact with tubulin, which is the protein involved in mitosis, so it causes chromosome loss and has other effects on mitosis. Bacteria don't undergo mitosis, so the bacterial system isn't going to pick up a chemical like griseofulvin.

There is now a whole battery of short-term tests, mainly looking at mutagenicity, and together they make a very powerful toxicological tool. In a number of cases, many of these short-term tests give the same answer, but I think the Salmonella test is never going to pick up every carcinogen. In addition there are some very important chemicals that Salmonella isn't picking up. One class is the glycosides. These are chemicals like cycasin, a beta glucoside of methylazoxy methanol, which is an alkylating agent. But once you have the glucose on it, it doesn't alkylate; it causes cancer in animals, but not in germ-free animals, and the reason is that the bacteria in the gut split off the glucose and liberate the methylazoxy methanol, which alkylates the animal's DNA and causes cancer. Salmonella doesn't split off the glucose, and liver doesn't split off the glucose; but if you add beta glucosidase, then it works.

We've been trying to develop a model for the human gut, because there are a number of these kinds of glycosides in nature where the glycoside isn't a mutagen, but if you split off the sugar, then it is mutagenic. Recently we've developed a model that consists of what we call fecalase. Human feces are half bacteria. If you add a little buffer, sonicate it, and make an enzyme extract, it has all the glucuronidase, glucosidase, mannosidase--all these kinds of enzymes. You can just add a little to the petri plate and it's sort of the equivalent of what the human gut is doing. And with that we can detect a few more of these chemicals.

In addition, there is a very important class of heavily chlorinated chemicals, and I think the most serious deficiency of the test system is that these are not detected as mutagens. There are chemicals like PCBs and PBBs and dieldrin and DDT and carbon tetrachloride. These don't work in the test system and yet they are carcinogens. Carbon tetrachloride is known to be metabolized to a free radical that hits DNA. We're working on that problem, and I think we may be able to crack it. It may have to do with short half lives of active metabolites.

There have been two other big validation studies, by ICI and by Sugimura in Japan, and they both got approximately 90%, but obviously that figure depends on what group of chemicals you think is representative.

Now what about noncarcinogens? That's a much stickier area because a carcinogen is something that causes cancer with 95% probability in a rat test or in people, but a noncarcinogen may be something that somebody tested in three monkeys or five rats, kept them for a year, and then said, "Oh, it doesn't cause cancer." So one really wants to know what the power of the test is when somebody says something is a noncarcinogen.

We looked at many close relatives of carcinogens and many common biochemicals, and almost all of them were nonmutagens but there were some mutagens. I'd like to make the argument that I think the problem here is mostly in inadequate cancer tests, and I'll give two examples of that.

One case we looked at was the 2-acetyl aminofluorene series. This was developed as an insecticide and someone did an animal cancer test and found it to be a potent carcinogen. So it's been used as sort of a model carcinogen over the years, and in our test system it produces 108 revertant colonies per nanomole, so it's quite a good mutagen. The 1-hydroxy derivative, the 3-hydroxy, 5-hydroxy, and 7-hydroxy and 4-acetyl aminofluorene have all been reported to be noncarcinogens. We looked at the mutagenicity, and 1-hydroxy-2-acetyl aminofluorene gave less than 0.02, 3-hydroxy less than 0.02, and

5-hydroxy less than 0.04 revertant colonies per nanomole. But for 7-hydroxy we found 0.42, so it was 250 times less active than acetyl aminofluorene when you put a hydroxy group on the 7-position. But we called it a false positive because, although it was clearly a mutagen and we could get a reproducible dose-response curve, we knew that the animal cancer test couldn't have picked up something even one-fiftieth as active as acetyl aminofluorene. But then Vic Donahue in the lab was doing high-pressure liquid chromatography, and he came in and said, "Aha, it's all due to an impurity." In 7-hydroxy-2-acetyl aminofluorene he found a tiny amount of acetyl aminofluorene, and when we looked at the clean 7-hydroxy it was negative. In fact they had done the cancer test on material that had a known carcinogen on it; they just couldn't detect it. And 4-acetyl aminofluorene is the isomer and we get a value of 0.3, and that's one of our so-called false positives because even though we find it 360 times less active than its isomer, the cancer test was negative, and therefore we call it a mistake--a false positive. So I think the problem is mainly the sensitivity of cancer tests.

In Japan furylfuramide was the main food additive that 100 million Japanese ate. It's a nitrofuran and it was added to the tofu, which is a bean curd that everybody in Japan eats. They used to have a very efficient distribution system for it, and people would buy it within a few days. Then they added the bacterial agent, so now it will keep for weeks. They had done a cancer test on this, and even though most nitrofurans are carcinogens, this one was negative in an animal cancer test. Recently some Japanese scientists found it was positive in E. coli, and it was a supermutagen in our system and a mutagen in yeast and a mutagen in Neurospora and a mutagen in silkworm, a mutagen in lymphocytes in culture, rat bone marrow cells. You could feed it to pregnant hamsters and mutate the cells in the embryo. So Sugimura and other scientists went to the government and said, "Maybe we ought to ban this even though it's not a carcinogen. We're a little nervous about all this." And the government said, "Well, it's very important economically and we don't know what all these tests mean. You go do some more cancer tests." So they did some more cancer tests and spent another year and a half doing that and now it has caused cancer in rats, mice, and hamsters. They just did more thorough tests, and now it's banned.

One of the other advantages of these short-term tests is that you can look at complex mixtures. Cigarette smoke condensate is an example. We get about 30,000 revertant colonies per cigarette or per 20 milligrams of tar; one can look at all the different fractions that the tobacco industry has fractionated cigarette tar into and they've isolated 2000 different chemicals from cigarette smoke condensate. They still have thousands more to go--it's a very complicated mixture. But the industry thought, "Aha, now we have a quick bioassay. We'll identify the mutagen and devise a filter and eliminate it." But when they started using the bioassay, they found hundreds and hundreds of mutagens, and they're not going to devise a filter, so maybe it's better just to give up smoking.

Edith Yamasaki and I recently worked out a way of putting human urine on our petri plates by concentrating the nonpolar chemicals on an XAD column first, and we could show that cigarette smokers had mutagens in their urine, but we didn't see any significant amount in the nonsmokers we looked at. It's clear that cigarette smoke actually is a fairly weak mutagen or carcinogen on a weight basis, but you're getting a gram of it in your lungs every day with two packs a day. When you're doing that for 30 or 40 years, it turns out that two packs a day is about eight years off your life. These chemicals are getting into you, giving you more heart disease, showing up in your urine, causing more bladder cancer, and now there's more and more evidence about genetic defects in children of smokers. Even a weak substance may be fairly dangerous when the dose is so enormous.

Also, one can look at many of the chemicals in the environment. I teach an undergraduate lab in Berkeley, and I always throw in one mutagenicity experiment. Berkeley students usually bring in marijuana or birth control pills or something else they're interested in. One year one of the students brought in his girlfriend's hair dye, which was mutagenic. Edie Yamasaki and I then showed that 90% of hair dyes were mutagenic. It turns out that when you dye your hair you can put four grams of aromatic amines on your head, plus a lot of hydrogen peroxide. A good number of the ingredients turn out to be aromatic amines that are mutagens, and a number of the others become mutagens when you mix them with hydrogen peroxide. Since then there's been a lot of agitation about all of this. The battery of short-term tests is coming up with the same general answer: many of these chemicals are mutagens.

If you plot production in hair dye use, almost 40% of women in the U.S. have dyed their hair and there are millions of regular hair-dye users. Again, whatever the effect of this is, we'll see, starting about now.

Tris [tris-(2,3-dibromopropyl)phosphate] was a flame-retardant chemical added to polyester pajamas; until recently, 50 million American children walked around wearing this material. About 5% of the weight of the pajama was this chemical just added to the fabric, not covalently reacted with it. Tris is insoluble in water, but it will dissolve in skin oil, and so there was an enormous surface being exposed. Tris turned out to be a mutagen. It's very similar to dibromochloropropane (another dibromo compound that is a mutagen), which sterilized about 100 factory workers in California, and to ethylene dibromide, a mutagen and carcinogen. Dibromopropanol is a metabolic product of Tris.

This work was done by Prival and Rosenkranz, using our test system, and Arlene Blum and myself. In our paper in Science, Arlene Blum and I pointed out that they even had an impurity of a known carcinogen (DBCP) in the Tris. So again one can use a chemical on a massive scale with really minimum studies on its toxicology, where even just by looking at its structure you would know it is a dangerous sort of chemical.

About 100 million pounds of flame retardants are made every year in the U.S. In Michigan, 500 pounds of polybrominated biphenyl, which was made as a flame retardant, got mixed in with cattle feed by mistake and caused a disaster in the state agriculture. This was recently shown to be a carcinogen, just like the PCBs. But this was just 500 pounds. Millions of pounds have been made, and will end up in the environment and bioconcentrate in animals.

Soon after we showed that Tris was a mutagen, and Prival and Rosenkranz showed it, we sent it around to all our friends, and it turned out to be a potent mutagen in Drosophila and was found to cause unscheduled DNA synthesis in human cells and sister chromatid exchange. Now the cancer tests have come in, and it's quite a good carcinogen in rats and mice and also has been shown to be a carcinogen in skin-painting studies in animals. All of these dibromo compounds are sterilants and cause testicular atrophy as well; DBCP caused infertility in people, and Tris has been shown to cause such problems in rats and mice as do ethylene dibromide and DBCP.

Recently we collaborated with Horning and Dougherty and showed that the urine of children who were wearing Tris-treated pajamas contains dibromopropanol, which is a hydrolytic product of Tris and is a mutagen itself. So it's clear that the substance is getting through human skin. The amounts are small and again one doesn't really quite know what the risk is, but it is on a massive scale.

If you drive in California in the cotton areas, that smell in the air is toxaphene. Toxaphene is probably going to be the last of the first generation of pesticides to be banned. It's the largest volume insecticide in the U.S., and it's been used in incredible amounts. It is made from the residue from making turpentine, from pine oil. You heavily chlorinate this residue and get a lot of chlorinated terpenes. Casida at Berkeley has shown that there are at least 177

28

different isomers in there, and what's so nice about this as an insecticide is that you never find it in mother's milk and you never find it in body fat because it's 177 different chemicals. But actually it bioconcentrates quite well and it's a mutagen (this is the work of Kim Hooper and myself), and it has also recently been shown to be a carcinogen and all sorts of other nasty things.

Whenever you chlorinate material that has organic matter, you make carcinogens. Sixty-six different chlorinated chemicals have been identified in Cincinnati drinking water, of which many are known carcinogens. If you chlorinate water without much organic material it really isn't so bad, but the more organic material, the more chlorouracil you make, and the more chloroform, and the more carbon tetrachloride, and so on. We use an enormous amount of chlorine in the U.S. to treat waste water as well, which is just filled with organic material. In addition, in paper bleaching the first step is to chlorinate all the organic material so that you make a whole zoo of chlorinated chemicals, which gets into the rivers and the lakes, and the fish bioconcentrate it and it gets into people. So more and more now people are looking for alternatives to chlorinating paper pulp, maybe using sodium hydroxide and oxygen as the first step. And also now, more and more people are starting to think about running water with organic material through carbon filters before you chlorinate it.

All of this sounds as if it's really the industrial world that's doing us in, but in fact the industrial world hasn't hit us yet. Most of the cancer now is probably of natural origin--cigarette smoke, ultraviolet light, and items in our diet, because most of the chemicals we get into us are in our diet. The hippies think that nature is benign, but it isn't.

Many studies have been done showing a correlation between fat intake and colon and breast cancer and heart disease. Colon and breast cancer are the most common forms of cancer after lung cancer. Right now there is a very interesting experiment being done by Bob Bruce in Canada. He has examined the feces of patients coming into the hospital with colon cancer, using our test system to see if he could find a mutagen. He has found a potent mutagen. The structure is not yet fully characterized, but it turns out to have oleic acid in it and some nitrogen. It has a nitroso group, so it looks like a nitrosamine. The whole class of nitrosamines are supercarcinogens. The interesting thing is that if he feeds a lot of vitamin C or a lot of vitamin E to these patients the level of the mutagen drops dramatically. It is known from animal experiments that if you feed nitrite plus a secondary amine you get lots of tumors, but if you feed a lot of vitamin C as well, you don't, because the vitamin C interacts with the nitrite. So maybe Pauling is right after all. There is a lot more to be done before this is proven, but I think it shows one of the uses of the short-term test.

Sugimura in Japan, who is one of the leading workers in cancer research, was watching his wife cook some fish on a hibachi one day. In Japan they often roast these little fish. He scraped off the surface to see whether it was mutagenic, and it was horribly mutagenic. So he said, "Well, what's in there?" So he heated up carbohydrate and protein and he found that if you heat up protein you get enormous numbers of revertant colonies if you put a little bit on the petri plate, and the more you heat it the more colonies you get. When you cook meat or anything with protein in it and char it a bit, like a charcoal-broiled steak, you're making mutagens.

Sugimura has now heated up kilos of tryptophan and glutamic acid, and so on, and has isolated many mutagens. He now has two of them that he's characterized from tryptophan, and they both work in another short-term test, transformation of animal cells in tissue culture. Now he's doing animal-cancer tests on them. Similarly, he's found mutagens, again using our test as a bioassay, produced from glutamic acid, and he's running transformation tests on them. One doesn't know whether a charcoal-broiled steak is equivalent to a couple of packs of cigarettes or what, but it's clear we're getting lots of mutagens from charring protein.

There are also nitrosamine carcinogens in our natural environment. We get nitrate into us and then convert it to nitrite that reacts with secondary amines to make nitrosamines. Mold toxins like aflatoxin, are powerful carcinogens, and many more are being discovered. Plants aren't benign either. Each species of plant is only eaten by a few kinds of insects. Plants and insects have evolved together, and plants have devised nasty chemicals through millions of years to kill off all the insects and animals chewing on them.

Quercetin is a very common flavonol that's present in practically everything we eat. We eat a gram of flavonoids a day and they are present as glycosides. The glycosides aren't mutagens, but several groups have shown that quercetin is a mutagen in our test system, and now it's also been shown to work in transformation. If you split off the sugar, then you get a potent mutagen. It's clear that the bugs in the gut have all the enzymes to split off these sugars, but it's not clear how much of the quercetin is absorbed, and how much of a hazard this is. What is clear is that there are many chemicals out there that are mutagens, and we're just going to have to spend a lot of time trying to figure out what's important and what isn't important--but it isn't that everything artificial is bad and everything natural is good.

I don't think the situation is hopeless, because at the University of California at Davis they spent 20 years breeding a tomato that is cylindrical, has a thick skin so you can drop it from 2 meters and it bounces, and it's completely tasteless. So if they can spend 20 years breeding a tasteless tomato, they ought to be able to breed the quercetin (or whatever) out of tomatoes if it is in there and if they need to. I think all of the plants we eat are highly bred, and we can deal with that once we know what's important and what's not important. That's what we have to do, and it's going to take a long time to figure out how to set priorities, but we're not going to ban all the mutagens and carcinogens.

In the last year or so we've been trying to analyze animal cancer tests for potency. We got into this partly because people kept saying, "Well, this substance is not a carcinogen and yet it's a mutagen." And we'd look at the cancer test and find it was a test that had been done on five animals or something like that. Also, Matt Meselson looked at our potency list for mutagens in our test system and said, "Well, how did that relate to potency in animal cancer tests?" He saw a rough relation between potency in Salmonella and potency in an animal cancer test on a few chemicals.

We decided to look at every cancer test in the literature that's suitable for calculation, to calculate the daily dose needed to give half of the animals cancer, and to look at a negative cancer test and estimate the power of the test. So now we have an enormous computer printout that contains most of the cancer tests in the world, which Hooper, Sawyer, Friedman, and I have been working on the last few years, with help from Richard Peto on the statistical problem. Saccharin shows up at about 5 grams per kilo per day to give half of the rats cancer; DBCP at about a milligram per kilo per day; aflatoxin in rats about a microgram per kilo per day; and TCDD still lower. So there's an enormous range of a hundred millionfold in potency of carcinogens.

Now what we want to see is how often rats are like mice, and obviously, for human exposure, one wants to know not only how potent a carcinogen is, but are species similar--can we make the jump from rats and mice to people? Priorities must take into account not only the potency of a carcinogen, but also human exposure. We might have a potent carcinogen inside a pipeline, and no one would care, but even a weak carcinogen like tobacco smoke may be very important if you get a lot of it into you and there are a lot of people exposed. So really both aspects come into play.

In this way, we've been analyzing the cancer data on the pesticide dibromochloropropane, which sterilized a number of factory workers. The data show a good dose response, in both sexes of rats and mice: A few milligrams per kilo per day for a lifetime gives half of the rats and mice cancer. The factory workers were breathing in nearly this amount. It turns out that just a few parts per million in air is a few milligrams per kilo per factory worker per day, because you breathe in 7000 liters of air a day, so that very roughly one part per million in the air is a milligram per kilo per worker per day. So in fact with many of these volatile solvents, such as ethylene dichloride, people are breathing in roughly the same amount, on a milligram per kilo basis, that is giving half the animals cancer.

Now let's say we can measure potency in animal cancer tests and get some priority list, but the number of chemicals in that list is very small. There are about 500 chemicals altogether for which we can get a rough idea of potency. But there are 50,000 chemicals in commerce, and there are many complex mixtures as well. We'd like to have animal cancer tests on all of these but we don't and won't. Yet we have to try to set priorities, since we're not going to ban everything, and with these enormous numbers of chemicals our DNA is getting hit from every direction.

So how do we set priorities? What we're exploring now is the possibility of a relation between potency in Salmonella and the other short-term tests and potency in animal cancer tests. It's clear that this is going to be a first approximation. Ground-up rat liver or ground-up human autopsy liver is only a first approximation of a person, and bacteria are just a first approximation of DNA. Is this ever going to give us a rough idea of potency? Matt Meselson and I think it looks promising, but it drives most toxicologists up the wall because they are used to monkeys or rabbits, and the idea that ground-up liver plus bacteria is going to say anything about potency of carcinogens makes them shudder; but if one needs just an order-of-magnitude idea, maybe one can get it this way. We will see.

To conclude, then, I think one could make a good case that, from the time we're born to the time we die, our DNA is getting hit by chemicals and by radiation and that's contributing to germ-line mutations, to the sizeable number of genetic abnormalities we have in the population, to cancer; and, also, I think you could make a good argument that it's contributing to aging and to heart disease, as well. Benditt up in Seattle has shown that atherosclerotic plaques are derived from single cells, and he thinks they may be like benign tumors. So heart disease may be mutational, and actually there's much more heart disease in smokers. All that will have to be worked out.

It is clear that cells have elaborate systems for protecting their DNA: repair systems that creep along the DNA trying to repair it, and other mechanisms. So DNA repair seems to be very important, and clearly DNA is important. Now we're going to have to study all those chemicals that are damaging our DNA, find out which are the more important ones, and find out how to minimize human exposure to them.

[A more detailed version of this talk (with references) has appeared in B. N. Ames, "Identifying Environmental Chemicals Causing Mutations and Cancer," Science, 204: 587-593 (May 11, 1979).]

Selected Readings

Hooper, N. K., B. N. Ames, M. A. Saleh and J. E. Casida. Toxaphene, a complex mixture of polychloroterpenes and a major insecticide, is mutagenic. Science 205:591-593 (1979).
Yamasaki, E. and B. N. Ames. Concentration of mutagens from urine by adsorption with the nonpolar resin XAD-2: Cigarette smokers have mutagenic urine. Proceedings of the National Academy of Sciences, Washington, 74:3555-3559 (1977).

HEREDITY AND CANCER IN MAN

Alfred G. Knudson, Jr., Director

Institute for Cancer Research, Philadelphia

Surely all of us who have returned to Caltech on this occasion can be forgiven for reflecting upon the effects of our first encounters with the Division of Biology. For me, this encounter with Morgan's Division was an introduction to living things of land and sea, to their development, and to their genetic variation. It is hardly surprising that pediatrics became my medical specialty and that years later I would want to return to Beadle's Division to try to "catch up" on the revolution in biology, a revolution that has had such great impact upon our knowledge of the subject of cancer.

For many physicians and scientists, the study of cancer is the study of life itself. Indeed, it is difficult not to be intrigued by a process that brings together the forces of nature and nurture upon the behavior of cells and their molecules to produce a major, and perhaps the most frightening, affliction of man. Yet despite our intense interest in, and the endless volumes written about, cancer, we seem to move so slowly toward its control. Over the 50 years that Biology has lived here at Caltech, neither the incidences nor case fatality rates have changed much for most cancers. But, thanks in large part to the speakers who have preceded me today, our knowledge of the beast, and the prospects for its conquest, have increased remarkably.

The study of cancer in man himself has also progressed considerably, despite the serious limitations that are inherent. Much has been learned about the causation of cancer by the study of its incidence as a function of time, place, and circumstance of life. Between the eighteenth century, when Potts first described a remarkable incidence of scrotal cancer in chimney sweeps, and the present, we have been provided with many examples of occupational cancer.

During this century, in the United States, although the mortality rates of most cancers have changed little, there has been an alarming increase in lung cancer mortality as a result of cigarette smoking, a significant decrease in uterine cancer mortality for several reasons, and a substantial decrease in stomach cancer mortality for no known reason. We find, too, a considerable geographic difference in mortality--sometimes explicable, as in the case of melanoma in the sun belt, sometimes not, as with colon and breast cancer in the northeastern USA. For only a few cancers, e.g., pancreatic cancer, do we find no meaningful geographic variation. A great contribution to our understanding of the possible meaning of geographic variation has been the study of Japanese migrants to this country, which has revealed in them, but more especially in their offspring, a sharp shift from the high incidence of stomach cancer and low incidences of colon and breast cancer that are characteristic of Japan, to the opposite incidences that characterize the United States. The differences between our nations seem to lie in environmental rather than genetic differences. It is from observations such as these that the conclusion has been reached by some that 80% or so of cancer is of environmental origin and preventable. For some of these cancers, the environmental factors are known; for others, they are not.

Those agents that have been shown to be carcinogenic, or are highly suspect, fall into the same classes as do those known to cause cancer experimentally in animals: irradiation, chemicals, and viruses. The observations that these same carcinogenic agents are also often, if not always, mutagenic, or in some way able to alter the host genome, have provided strong support for the notion that cancer arises by somatic mutation and that cancer is a genetic disease of somatic cells. In proposing such a mechanism in 1914,

32

Boveri also noted that cancer should be clonal in origin, i.e., arise from single cells. The study of human cancers in females heterozygous at the X-linked glucose-6-phosphate dehydrogenase locus has fulfilled that prediction.

But the origin of cancer in man is not exclusively environmental. It has long been known that some individuals are genetically predisposed to cancer. Boveri took special note of xeroderma pigmentosum, a recessively inherited disorder that is invariably associated with skin cancer. This disease has been the subject of intensive study since Cleaver's discovery a decade ago that cells from these subjects do not show normal repair of ultraviolet light-imposed damage to DNA. Such cells also mutate at increased rates and therefore provide considerable support for the idea that ultraviolet light causes cancer via mutation.

One of the major problems in human carcinogenesis is the assessment of the interaction of heredity and environment. It is even possible that most human cancer results from the action of environmental agents upon genetically susceptible hosts. But there are also numerous dominantly inherited cancers that seem to be determined solely by genetic endowment, there being no evidence that environmental factors are operating. These germinal mutations seem to replace the somatic mutations induced by environmental agents.

Although environmental and/or genetic factors can be identified for many cancers, neither seems to play a role in some instances. For some cancers, there is a rather constant incidence over time and place, as if "spontaneous," universal forces are operating. According to a somatic mutation hypothesis, these cancers result from background, or spontaneous, mutations, as distinguished from induced mutations. The forces of nature and nurture therefore seem to be disposed in all four possible combinations of one, the other, both, or neither. The incidence of the "background" cancers would seem to be some lower limit below which the incidence of cancer cannot fall, but at the same time an incidence toward which we should strive with prevention and which we could hope to control with treatment.

Recognizing that heredity and environment are both major variables in the origin of cancer, we now turn to the most important variable of all--age. It is well known that the incidence of most cancers rises sharply with age. For cancer, generally, this rise is nearly logarithmic. For some particular cancers, the rate of increase declines in later years, for some it increases. Several cancers, notably colon and stomach cancer, increase with a power of age. For these two the power of age for incidence (I) is approximately six; for accumulated incidence, or prevalence, at a given age (P), one more than that, seven--i.e., $P = k_1 t^r$, $I = k_2 t^{r-1}$, where r = 7 for these two cancers.

An explanation for this relationship between cancer incidence and age was first offered in 1951 by Nordling, who proposed that $\underline{r}$ mutations must occur in a cell before it is transformed into a tumor cell. Assumptions implied in this hypothesis are that the number of target cells does not change with time, that there is no differential growth of cells that have not sustained the last mutations, and that mutation rates are constant with age.

Armitage and Doll later suggested that the number of mutations could be as few as two, if singly mutant cells had a growth advantage over normal cells. Fisher proposed a modification of this hypothesis that allowed for a smaller growth advantage of intermediate cells and a total of three mutations. Fisher noted that, according to his hypothesis, one mutation would produce cancer with a value of r = 1, two mutations, r = 4, and three mutations, r = 7, according to an equation r-1 = 3 (n-1), where $\underline{n}$ is the number of mutations required.

A direct test of the multiple mutation hypothesis for the origin of cancer is not possible at this time. However, there is one indirect test made possible by a comparison of the hereditary and nonhereditary forms of the same cancer, a comparison that is possible for colon cancer because there are appropriate data on subjects with the dominantly inherited condition, polyposis of the colon.

Here we can suppose that one mutation of the n mutations has been inherited and that the age-specific incidence and prevalence curves will change accordingly to some value r'. If r = n, as originally proposed by Nordling, then r' should equal six. If Fisher were correct, r' should be four. In fact r' = 4, suggesting that n = 3.

This is a particularly interesting finding because it is in accord with the observation that the colonic mucosa of polyposis subjects remains normal in appearance in many areas (containing the one inherited mutation), develops adenomatous polyps in some areas (two mutations), and progresses from these to carcinoma in a few areas (three mutations). The third "mutation" may be a critical step in a sequence of changes associated with the evolution of new karyotypes, since 3-step tumors are always chromosomally abnormal, whereas 2-step tumors such as the adenomatous polyp may not be.

Related to the influence of age upon the incidence of cancer is the phenomenon of latency. It was observed even by Potts that cancer typically developed after an interval of 10 years or so following exposure. Many such examples are now known, although the duration of the latent period may be as brief as a year or two for some childhood cancers following irradiation in utero or as long as 30 years or so for mesothelioma following exposure to asbestos. Presumably, the environmental agent produces a mutation in a cell whose descendants are at risk of further mutations, spontaneous or induced. Cancer develops when the critical number of mutations is reached. The length of the latent period should be a function of the number of target cells, the rates at which mutation occurs per cell division, the frequency of cell division, and the number of critical mutations.

In subjects who inherit a cancer mutation, the latent period should be shortened. This is certainly true for all of the dominantly inherited cancers; the age-specific incidence is shifted to earlier ages, as observed already for polyposis of the colon. One dominantly inherited condition, the nevoid basal cell carcinoma syndrome, is especially interesting in this respect. Skin cancer develops in such subjects in the second decade of life under normal circumstances. But some of them also develop a brain tumor, medulloblastoma, for which X-irradiation is administered to the cranium and spinal region. In patients without the syndrome who are so treated, skin cancer can be observed in the overlying skin 15-20 years later, presumably because the radiation produces a somatic mutation that is similar to the germinal mutation in the syndrome. In the syndrome patients, however, the radiation would be mutating cells that are already mutant. Strong has observed that in such cases skin cancer can be observed in the field of irradiation in less than one year, suggesting that irradiation has supplied a final rather than initial mutation in them.

For most cancers the pathway from normal cell to malignant cell is tortuous and improbable of attainment. Two or more rare events seem to be necessary for the production of the first cancer cell, and then its growth to a clinically significant tumor must ensue. Genetic and environmental factors may hasten the process by reducing the number of steps or increasing the rate at which they occur, so the finally attained incidence of a cancer reflects spontaneous processes in germinal and somatic cells and the modification of these processes by environmental agents.

What, then, do we know about the number and nature of the genes whose modifications can lead to cancer? To gain some further insight we can turn to the cancers of children. Although these tumors are uncommon, their rapid growth, easy detection, and occurrence in both hereditary and nonhereditary forms render them particularly suitable for analysis. Especially useful is retinoblastoma, because its treatment has been so successful that many patients have survived to adulthood and reproduced. What is often observed is that such survivors have affected children, even when their own parents and other ancestors have not been affected. The frequency of such new genetic cases can be attributed to new germinal mutations occurring at a retinoblastoma locus at a rate of about 6 mutations per million gametes. About 25% of cases are bilaterally affected. Enough such patients have now survived to inform us that 50% of their offspring are at risk of retinoblastoma. On the other hand, only 10–15% of the offspring of unilaterally affected cases are affected. The explanation of this discrepancy is not that there is a "bilateral gene" and a "unilateral gene," because the offspring of bilateral cases may be unilaterally affected, and the affected offspring of unilateral cases are usually bilaterally affected. A better explanation is that 40% of all cases are "genetic," i.e., associated with a dominantly heritable germinal mutation and that 60% are "non-genetic." Bilateral cases all belong to the former category, unilateral cases to both categories.

From an analysis of hereditary cases of retinoblastoma, it can be readily concluded that most individuals who carry a germinal mutation for retinoblastoma develop more than one tumor. Both indirect consideration of the ratio of unilateral and bilateral cases among gene carriers and direct counting of tumors in eyes lead to the conclusion that a mean number of 3–4 tumors develops per gene carrier. On the other hand, the incidence of non-genetic cases is about 3 per 100,000 children, so the relative risk of tumor imposed by the germinal mutation is approximately 100,000.

Even though most gene carriers develop at least one tumor, the probability of such an outcome is low when viewed at the level of the cell. There are millions of mutant retinoblasts in the developing retina, yet only a few are transformed to tumor. Evidently this mutation is not sufficient by itself to produce tumor; in fact, the probability of such tumor seems to be of the order of magnitude of 10^{-6}. A simple explanation is that a second mutation must occur before transformation can occur. In both genetic and non-genetic cases, two mutations are necessary; one is germinal and one somatic in the first instance, both somatic in the second.

A direct test of this model is not possible but an indirect test has been made by comparison of age-specific incidence data with the expectations of a mathematical model. The model begins by noting the similarity in the shapes of the age-specific incidence curves for embryonic tumors of children and for the populations of embryonic stem cells; both have peaks in early childhood. Another way of recording these changes is to note the manner in which patients who ultimately develop tumor do so as a function of age. The still undiagnosed cases disappear faster for hereditary cases than for nonhereditary cases, and, among the former, bilateral cases become apparent earlier than do unilateral cases. A two-mutation model supposes that the second mutation must occur before differentiation is complete and cell division ceases. A comparison of expected and observed incidences reveals a close fit. In addition, the necessary somatic mutation rates have been calculated to be in the range 10^{-6}–10^{-7} per locus per cell division, values not very different from those obtained for human cells in vitro. The total incidence of such an embryonal tumor can then be given as a function of germinal and somatic mutation rates (μ_g, μ, and ν), the coefficient of selection of the germinal mutation(s), the initial number of stem cells present [b(o)], and the total number of cell divisions of the target cells [a(∞)]:

$$\text{incidence} = \frac{2\mu_g \, (1-3^{-\mu \cdot a(\infty)})}{s}$$

$$+ \, \mu \cdot \nu \cdot a(\infty) \cdot \ln\{\frac{a(\infty)}{b(o)} - 1\}.$$

If environmental mutagens are distributed unevenly, there should be geographic differences in tumor incidence. For each of two embryonal tumors, retinoblastoma and Wilms' tumor of the kidney, the incidence is remarkably constant worldwide, suggesting that germinal and somatic mutations are occurring at background rates for them.

If two mutations are necessary for tumor production, what is their relationship to each other and what is the normal function of the genes involved? The answers are not available, but there are some clues. Cytogenetic studies of retinoblastoma tumors demonstrate only one frequently observed abnormality, a deleted or absent chromsome 13. This observation has been made in both hereditary and nonhereditary cases whose somatic cells show no abnormality. In addition, there are some rare patients with retinoblastoma whose somatic cells all show a deleted segment of the long arm of chromosome 13. Such subjects have the same high risk of tumor that mutant gene carriers have. Finally, some recent studies of families in which an affected parent shows dimorphism for the short arm of this chromosome reveal a strongly nonrandom segregation of the two in the unaffected and affected children. It seems to be a safe conclusion that all cases of retinoblastoma have a mutation in at least one chromosome 13, that in hereditary cases the abnormality may be either visible or not, and that in nonhereditary cases the same change occurs somatically. Similar data are not available for other tumors, although several recent reports demonstrate that in some patients with Wilms' tumor all somatic cells contain a deletion of the short arm of chromosome 11.

Although there are no other instances of specific prezygotic chromosomal abnormality associated with a specific tumor, there are increasing numbers of instances of what seems to be specificity of somatic chromosomal abnormality. The best known example is that of the Philadelphia chromosome, a small chromosome 22 whose deleted segment has been translocated to another chromosome. This abnormality is specifically associated with chronic myelocytic leukemia. Other abnormalities are associated with other types of leukemia. Several forms of lymphoma, including Burkitt's lymphoma, thought by many to be of viral origin, are associated with abnormality of chromosome 14. Of special interest is the finding that abnormality of this chromosome has been found in the lymphoid cells, but not fibroblasts, of patients with ataxia telangiectasia (AT), a recessively inherited disease that predisposes to lymphoma. Chromosomal breakage is excessive in AT cells, and they are deficient in the repair of DNA damage inflicted by ionizing radiation. Although many types of chromosomal abnormality are produced, it is the specific change in chromosome 14 that imparts a selective advantage and results in the emergence over time of clones of neoplastic lymphoid cells. These neoplasia seem to be related to a specific chromosomal change that can be produced by more than one mechanism.

The site of a second mutation is not known. Considering retinoblastoma again, one possibility is that a second mutation occurs in the chromosome 13 that is already abnormal. This seems unlikely because some deletion cases show a deletion of almost the entire long arm. A particularly intriguing possibility is that the second mutation occurs at the same gene in the homologous chromosome. Thus, if the chromosome 13 can be defective by virtue of absence (13^-), visible deletion ($13q^-$), or invisible mutation (13rb), tumor could result from any of six combinations, all equivalent with respect to loss of a critical locus. At the cellular level, tumor formation would be a single recessive gene disorder.

If tumor results from homozygous loss of function, then another mechanism could be operating. Cells heterozygous at a critical tumor gene could become homozygous by somatic recombination. We do not know for many what the frequency of mitotic recombination of homologous chromosomes may be, and whether it can be more frequent than somatic mutation. It is possible that the second event is usually such a recombination rather than mutation. This mechanism could account for the action of certain chemical "promoters" of carcinogenesis that have no effect experimentally in the absence of mutagenic carcinogens. Similarly, any genetic condition that favored somatic recombination would predispose to neoplasia of all types, depending upon which genetic locus sustained a first somatic mutation. There is in fact one condition in man for which this may be the case--Bloom's syndrome. This recessively inherited disorder predisposes to many kinds of neoplasia. Somatic cells of patients show increased chromosomal breakage, a high rate of sister chromatid exchange, but, more importantly, homologous chromosome exchange. Of special interest, then, is the report in press from Radman in Belgium that certain chemical promoters greatly increase rates of sister chromatid exchange.

We come then finally to the question of the nature of genes whose mutations lead to cancer. Data on man cannot answer this, but there are some observations on dominantly heritable cancers that should be considered. For the most part, these mutations are specific for one or a few tumors. They not only demonstrate tissue specificity but even specificity within a tissue. For example, several such germinal mutations affect the nervous system, with specificity for retinoblastoma, neuroblastoma, medulloblastoma, or glioma. Of special interest are mutations that affect the adrenal medulla. At least one predisposes to neuroblastoma, while at least two others predispose to different forms of pheochromocytoma, a much more differentiated tumor. Evidently, tumor genes can be specific not only for a tissue, but for the level of differentiation within that tissue.

Clues to the action of tumor genes could come also from the study of the normal target tissue in hereditary cases. The best studied example is that of Wilms' tumor. In the nontumorous portions of the kidney of genetic cases, but not the nongenetic cases, abnormal clusters of poorly differentiated cells are found in the subcapsular areas of the cortex. Ultimately, these either differentiate normally or degenerate, leaving small scars. Some of the clusters resemble tiny Wilms' tumors. It appears that the Wilms' tumor gene interferes with, or delays normal differentiation.

What little evidence there is regarding the function of the normal alleles of tumor genes is compatible with the view that they are concerned with differentiation and that mutation arrests this differentiation at a specific stage in specific tissues, leaving the incompletely differentiated cell to continue in a mitotic cycle. Differentiation genes, human tumor genes, and the genes that regulate tumor virus expression may be identical.

Selected Readings

Hethcote, H. W. and A. G. Knudson. A model for the incidence of embryonal cancers: application to retinoblastoma. Proceedings of the National Academy of Sciences, Washington, 75:2453-2457 (1978).

Kinsella, A. R. and M. Radman. Tumor promoter induces sister chromatid exchanges: relevance to mechanisms of carcinogenesis. Proceedings of the National Academy of Sciences, Washington, 75:6149-6153 (1978).

Knudson, A. G. Genetics and etiology of human cancer. Advances in Human Genetics 8:1-66 (1977).

Knudson, A. G. Retinoblastoma: a prototypic hereditary neoplasm. Sem. Oncology 5:57-60 (1978).

Mulvihill, J. J., R. W. Miller and J. F. Fraumeni Jr., eds. Genetics of Human Cancer. Raven Press: New York (1977).

Schimke, R. N. Genetics and Cancer in Man. Churchill Livingstone: London (1978).

SESSION II: PHAGE

Introduction

Max Delbrück, Nobel Laureate

Board of Trustees Professor of Biology, Emeritus
Caltech

The idea for this little phage symposium originated on the day before Bill Wood's departure from Caltech. He and I were talking about the move, his move away from here to the University of Colorado, and he remarked that with his departure several decades of phage research in the Biology Division at Caltech came to an end. Sinsheimer's group had been dissolved a few months earlier with his departure to become Chancellor of the University of California at Santa Cruz. Now it is true that there is still phage research at Caltech--namely, Judy Campbell, who has a very interesting group in the Chemistry Division, not in Biology. But then the Chemistry and Biology Divisions are so tightly woven together that we can claim not to be quite denuded of phage research. In addition, of course, phage lambda plays a crucial technical role in the various recombinant DNA projects engaging the groups of Tom Maniatis, Eric Davidson, Lee Hood, James Bonner, etc.

Anyhow, while we were chatting about all this, we thought, shouldn't we have an Irish wake or something to commemorate the demise of phage research here, and how it might go on, in another world, in other labs; in other words what its future might be. So we started plotting this symposium, and after some back and forth it was decided to fuse it with the 50th Anniversary affair, the idea for which had originated and been discussed for a while in some other quarters. So a year and some months later, we had three speakers and a chairman to reminisce and to reflect, to inform and to amuse, and possibly, by informal talk, permit new ideas to be germinated.

I. Early Phage Research at Caltech

Phage research at Caltech was started in 1937 by Emory Ellis. He initiated phage research here in the following way: He had been trained as a chemical engineer, and jobs being hard to get, somehow he got a fellowship that Dr. Seeley Mudd had instituted to do cancer research. So, being a man who can do anything, he started to figure out what can you do in cancer research, and in reading about cancer he found that there was a virus angle to it, and in reading about viruses he found out that there are bacterial viruses--viruses that attack bacteria; and he thought they might be more manageable for an experimental approach, especially by a one-man team.

He started reading about bacteriophages, and then he persuaded Dr. Morgan to let him get the equipment together: an autoclave which was about the size of a pressure cooker, a little sterilizing oven, 40 pipettes, and about 40 petri plates. He went down to the University of Southern California, Department of Bacteriology, to his friend Carl Lindegren, and obtained from him an organism that hardly anybody in the Biology Division had heard of before: E. coli. Then he went down to the Los Angeles sewer department and procured one liter of sewage and filtered that and from that filtrate isolated a phage against the strain of E. coli. He also started studying the growth of this phage and made the first one-step growth curve.

At that time he gave a seminar on his work which I unfortunately missed. I had come in the fall and I had seen the seminar announcement on the calendar

39

and I had heard that there existed such a thing as phage, but I missed the seminar, having gone instead on a camping trip to Arizona and New Mexico with Frits Went. So I visited Ellis in his lab after I came back from this trip and he showed me plates with plaques, and after some back and forth he permitted me to join him. I had come here on a Rockefeller fellowship to learn Drosophila genetics, but I decided phage research looked much more attractive; you could do beautifully simple experiments one day and have the answer the next day or even sooner. So we had an absolutely marvelous time for a year. Then, unfortunately, Emory's sponsor discovered that this research was not really cancer research for which he had given his fellowship, and said, "You cut up mice or else." So Ellis had to go back to cutting up mice, but I continued.

Figure 1 shows two photos of Emory Ellis, taken in 1940 and in 1970. They are almost indistinguishable and, in fact, even in 1978 he still looks very much the same--exactly the same mustache, etc., the hair, too.

FIGURE 1

In 1969, when a Swedish television team came over here to make a movie for their Nobel celebration, we decided to fake a little slip of movie of Ellis and me doing an experiment together, and since he hadn't changed and I hadn't changed so very much we thought it might actually work out. In this little movie Ellis and I are doing an experiment in Room 01 Kerckhoff, and are getting our plates sorely contaminated because our sterilizing technique wasn't yet perfected. Especially we had a burette which we used for single burst experiments. This is a type of experiment in which you study the yield of phage from individual bacteria, which you isolate by making a high dilution of the culture and then incubating single drops in separate test tubes. Our autoclave wasn't big enough to sterilize this burette, so Emory thought it was good enough just to wash it out, but since we had from the previous experiment left the bacteria incubating there with phage, about 10^9/ml phage had developed, the washing out wasn't very successful, so from day to day our single bursts got bigger and bigger.

This was a very early period of phage research at Caltech, 1937-1939, and then there was a hiatus. As I said, Ellis had to drop out earlier and I went to Vanderbilt, and there and during summers in Cold Spring Harbor a number of people helped me to continue work on phage until I came back to Caltech in 1947 to be here on the faculty. From 1947 on, phage research continued in Biology at Caltech until last year. In 1948 the first one to come was Hershey, just for a few months, to help get the place warmed up, as it were. A little later, I think in the fall of 1948, Gunther Stent came. He had just gotten his PhD in chemical physics of macromolecules and knew as little about biology as I knew, and shortly afterwards came Elie Wollman from the Institut Pasteur in Paris, and Wolf Weidel from Butenandt's Max Planck Institut in Tübingen, Germany.

That was the first group. Stent, as you know, has made wonderful contributions to phage research but in recent years has switched to neurobiology. And he has also become a do-it-yourself philosopher like some of us and has written exceedingly interesting books and articles on the general relation of science, epistemology, and history of ideas, and some prognostications as to what might happen in the interaction between science and our society. Wollman went back to Paris to the Institut Pasteur after his stay here and got involved in bacterial genetics and has made magnificent contributions together especially with Jacob. A year later came Benzer and Dulbecco, and Jean Weigle turned up, and commuted. Weigle was here for many years, and he commuted between here and Geneva. He always spent about six or eight months here and then three months in Geneva, where he had been head of the physics department and professor of physics; he retained his professorship and went back and forth. It was a very good arrangement; he had an immense influence on all of us, and especially on some of the graduate students. Dulbecco came here as a senior research fellow, having grown up in Italy and having been with Luria for two years before he came here.

In this period we did a lot of work on, first of all, UV radiobiology of phage, which showed interesting effects which were uninterpretable and which only gradually and in part have become interpretable. I would now say, in retrospect, that this was not very profitable. Similarly we spent a great deal of time on the strange effects that Tom Anderson had first found that one of the phages could be activated by tryptophan, and I found that this could be counteracted by all kinds of conditions. These effects were studied in great kinetic detail, but this, too, was premature research because at that time it was quite impossible to guess what was really involved. It turned out much later that what was involved was something very subtle happening to the tail fibers of the phage.

The next group of people includes Peggy Lieb, Ole Maaløe from Copenhagen, Neville Symonds, Joe Bertani, who stayed for quite some time, and Gordon Sato, a graduate student and a real disaster kid at that time, and Dale Kaiser, also in 1950, and in 1953, just overlapping with the discovery of the double helix, was a short postdoctoral fellowship of Bob Sinsheimer.

The year after that, 1954, Jim Watson came; this was while I was away for a summer semester in Göttingen. I had left Jim in charge of the phage group, and that was a total disaster because Jim suffered from post partum depression after he had made his tremendous discovery. He was in very bad shape psychologically at that time and was absolutely useless as far as the group was concerned. He went from here to Harvard and he recovered, as you know, magnificently. Then Niels Jerne came for six months. He came from the Serum Institute of Copenhagen, where everybody has six technical assistants, and since he didn't have these here he decided it was a waste of time for him to do experiments and instead he wrote a fundamental paper on the details of the clonal selection in immunology. Naomi Franklin came at the same time and Walter Harm and Helga Harm, and shortly afterwards Frank Stahl came as the first of the graduate students of Gus Doermann from Rochester. Of course, Doermann spawned quite a few important graduate students: Frank Stahl and Bob Edgar and Dick Epstein.

The group Bob Edgar developed during this period was unique. He had a magnificent flock of graduate students, and also Charley Steinberg was there at that time and, part of the time, Dick Epstein. Meselson had been a graduate student in chemistry. He and Stahl, together with Vinograd, developed the density gradient centrifugation and applied this to prove the semi-conservative mode of replication of DNA, one of the fundamental experiments. Then in June 1960 a singular event occurred. Jacob came here from Paris and Brenner from England and together with Meselson they did the experiment showing that messenger RNA is transient and quite distinct from ribosomal RNA. There had been confusion for several years as to what the ribosomal RNA was doing and Jacob and Monod had a paper in press proposing the idea that messenger RNA was something distinct and was turning over rapidly. This prediction was proved with a T4 experiment in June 1960.

We have a very nice picture of all of these: Meselson, Jacob, Brenner, Gunther Stent, Charley Steinberg, in our garden. This photo has now been published in Horace Judson's book The Eighth Day of Creation. Around that time, Richard Feynman decided he also wanted to find out what this molecular genetics is all about, and he joined Bob Edgar's group and they did experiments on the strange interactions of several mutations in the same cistron. They almost but not quite discovered the idea of frameshift mutations. Feynman spent one whole summer and then the rest of the year he came over on weekends. He came in on Friday and ordered all the plates that he wanted and came on Saturday and did a splendid experiment and then came on Sunday and counted his plaques and evaluated his experiments, and then he came back next Friday and did the next experiment. I was totally amazed how he took to experimenting like a duck to water.

So this is a bird's eye view of the early period of phage research at Caltech. I would say the most significant part of it was the period after the molecular aspect had really gotten under way and when Bob Edgar had gotten his group together. Edgar, actually, had two great episodes: one, the conditional lethals which advanced so enormously the genetic fine structure analysis, and which helped to understand the genetic code termination signals. That was the one episode. The second episode is the one that Bill Wood is going to talk about: the assembly problems. This occurred after they got together in 1965 when Bill Wood came to Caltech.

II.   The Present Status and Future of Phage Research

The phages discussed at this meeting, ØX174, T4 and T7, all DNA phages, represent three classes that differ radically from each other. T4 belongs to the most complex class, the T-evens, with a very large number of genes and functions and an astoundingly complex morphology, so complex and sophisticated in fact that one is tempted to look upon it as having its evolutionary origin as an originally free living form that has become parasitic. This temptation must be resisted, however, for there is an enormous gulf between these phages and anything resembling a cellular organization. See the enlightened discussion in Chapter 19 ("Origin and Nature of Viruses") in General Biology by Luria, Darnell, Baltimore, and Campbell, 3rd edition, Wiley, 1978.

ØX by contrast belongs to the group of the smallest phages and was selected by Sinsheimer for that reason. He had started working on a relative of ØX, S13, but then switched to ØX because of its greater stability. Both of these phages, as well as several other representatives of this class, had originally been isolated in Paris in the early '30s by Sertic and Boulgakov. ØX was the first DNA phage that turned out to be single-stranded and a little later turned out to be circular, both astounding novelties at the time of discovery. And it has continued to yield astounding novelties, the most astounding being the multiple reading of its DNA message in different reading frames. This discovery is so new that one cannot yet assess its generality nor its evolutionary meaning. At first sight one is tempted to look upon the multiple reading as an economic device to save coding space but this argument is rather questionable since ØX seems to waste three major proteins to construct its capsid while other similarly small phages get along with a single one. We find that learning all the facts about ØX does not tell us the evolutionary reasons for these facts. If anything, learning the facts has greatly increased the number of questions one would like to ask from a broader perspective.

Phage T7 holds a middle ground in size and complexity and degree of completeness of factual information about its structure and functions. However, with T7 Bill Studier has made the first decisive inroads into an area that might be called comparative anatomy of the DNA. He is studying a number of related strains collected in the wild and laboratory strains derived from a common ancestor and he has given us the first inkling of the degree of constancy and variation to be encountered when strains are subcultured for decades in the laboratory. In the old days this was a nightmare we never dreamed we would be able to resolve: How do our strains vary from day to day, from year to year? How much subliminal variation creeps in that we do not know about? Bill Studier's experiments give a clear and definite answer to this question, or at least they outline a clear procedure for getting an answer. At the same time the experiments also make a beginning in the field of molecular evolution as applied to virology. Molecular evolution has been very successfully launched first by studying proteins in eukaryotes, then extending this study to prokaryotes and most recently extending such studies to various classes of RNA. For the first time we find here powerful and reliable indices of relatedness on the macro-evolutionary scale. Bill Studier's essay represents the first attempt to apply this methodology to viruses. It is my feeling that this procedure represents our only hope to obtain decisive insights on the evolution of viruses and their relation to the rest of evolution.

ØX174, A RESEARCH ODYSSEY: FROM PLAQUE TO PARTICLE, AND MUTANT TO MOLECULE

Robert L. Sinsheimer, Chancellor

University of California, Santa Cruz

Well, it's very nostalgic to stand here at this podium and to hear Max recount the glorious years, the glorious people. While it's true that phage work has, for the moment, migrated from Biology to Chemistry, I can't believe this is the last phage symposium that will be held at Caltech.

In that odyssey which is the record of the human exploration of nature, various fields go through golden ages. Each is a time when the combination of accumulated insight and available technique and research interest permit a rapid cumulative advance in the field, until it has been extensively mapped and the original problem solved, or it has been resolved to a deeper set of problems that must await a newer time. We can all identify such times in organic chemistry, in atomic structure research, in theoretical physics, and so forth.

Currently we are surely in a golden age of molecular biology. Although it is true that one of the speakers in another session of this celebration has earlier lamented the end of the golden age of molecular biology, most of us regard that eulogy as considerably premature. Some disciplines of molecular biology may of course thrive and wane as the wave of progress advances. Just what should be considered the vital signs of a golden age may be debatable, but out of curiosity I thought I would inquire into the health of that subfield of molecular biology known as ØX174 research. As the simplest of accrued indices I merely looked up the number of ØX174 references cited in the Chemical Abstracts Index per year (Figure 1). The graph presents the number of ØX papers per year in Chemical Abstracts as a function of time; obviously the field is not moribund nor indeed is there evidence of decline. Conceivably it is at its peak. Only the future will tell, but at the moment it is clearly a thriving industry.

Let me add that I do not regard this as a significant contribution to the sociology of science. A much more sophisticated and interesting analysis could be made of the changes in content and precision of discussion in these papers of which the total incidentally is a depressing 584. (That's about one per nine nucleotides. I am not sure what to make, of that.) While there is a certain historical cast to this symposium, I don't intend to drag you through these 584 papers; rather, my primary objective will be to try to present our understanding of the biology of ØX as we know it today.

While viruses of this type have undoubtedly been about for a very long time, they first appeared to the world of science as a plaque--a large plaque in fact--on a bacterial plate in the laboratory of Sertic and Bulgakov about 1934. The large size of the plaque suggested that the virus might be small. Filtration and radiation inactivation experiments supported that suggestion and indeed that thought was, as has been previously mentioned, the reason why we undertook to study this virus. The thought was that if it was a small virus we could hope to learn in considerably more detail the essence of its structure and function. I think that one can say now that hope has been fulfilled because it certainly is fair to say we know more about this particular virus than any other DNA-containing virus.

Hard evidence as to the size and other characteristics of the virus was not really obtained until it was purified and made available in large amounts beginning in the mid-1950's. Once it was purified it became evident from electron microscopy and from light-scattering studies that the virus was indeed small. Light-scattering data provided a particle mass of about 6.2 million daltons. Figure 2 is an electron micrograph of Fred Eiserling's which shows the comparative size of the ØX virus compared to a T-phage or to a tobacco mosaic virus. It is about 25 nanometers across. (As one excursion into history, I got out what I believe to be the first electron micrograph of ØX [Figure 3], which I took back at Iowa State before coming to Caltech. You can't see much more than a bunch of spheres, but at least you could measure the diameter.) Progressively better pictures were, of course, obtained by more expert electron microscopists. Figure 4 shows pictures with rather more detail, where one can begin to see the various knobs. Indeed you can see by extrapolation, assuming that there is another side to the virus, that this is indeed an icosadecahedron. Figure 5 shows pictures by Tromans and Horne, empty particles, indicating again, the icosahedral structure, and Figure 6 shows still more detail, getting down toward the individual subunits.

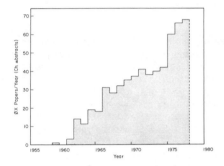

FIGURE 1 — *The number of articles about bacteriophage ØX174 cited in Chemical Abstracts in each year, 1957-1977.*

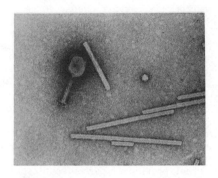

FIGURE 2 — *Electron micrograph illustrating the comparative sizes of bacteriophage ØX174, bacteriophage T2, and tobacco mosaic virus (courtesy of Prof. Eiserling).*

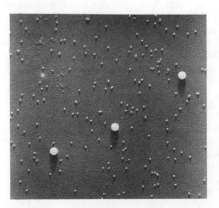

FIGURE 3 — *Early (1957) electron micrograph of bacteriophage ØX174.*

FIGURE 4 — *Later (1962) electron micrograph of bacteriophage ØX174 inducing knob-like projections (Maclean and Hall, J. Mol. Biol. 4, 1962).*

45

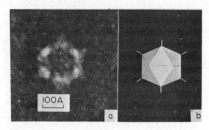

FIGURE 5 — *Electron micrograph of hollow particles of bacteriophage ØX174 illustrating the icosahedral structure (Tromans and Horne, Virology 15, 1961).*

FIGURE 6 — *Electron micrograph of bacteriophage ØX174 illustrating details of structure of the viral projections.*

We now know that the particle itself is composed of four proteins encasing the internal nucleic acid. There is the capsid protein of which there are 60 molecules, three per face of the icosahedron. This we know to be the product of cistron F. It is a protein of about 48,000 molecular weight. Then there are the two proteins that comprise the knobs or projections, of which there are 12. There is the cistron-G protein, about 20,000 molecular weight, of which there are five in each knob; and then the cistron-H protein of about 36,000 molecular weight, which presumably forms some sort of cap on the knob--there being one per knob. And then there is a small basic protein, cistron-J protein, of only 38 amino acids (12 of which are lysine and arginine), which is thought to be internal, providing a partial counterbalance to the charge of the nucleic acid.

Figure 7 shows a schematic drawing as to how we imagine, in some approximate way, the faces are composed, and then the knobs are built out at the vertices of fivefold symmetry. The knobs of course are essential for attachments to the host cell, and whether there is any differentiation among the knobs is still an undetermined question. Antibody inactivation studies would indicate that a single antibody is able to inactivate the particle, but of course that may reflect some distortion produced thereby. Pictures of the particle (Figure 8) as attached to the cell (one can see a great number of them) indicate that there are clearly many absorption sites. And then pictures looking down on an attached virus show that the particle attaches by one knob and is significantly embedded into the bacterial surface (to nearly halfway) in the process. Presumably, this process of attachment of the knob opens up a passageway for the internal nucleic acid to penetrate through and into the cell.

Now the DNA, which is inside, is, as is well known, a single-stranded ring, as is shown in these electron micrographs (Figure 9) taken by Dr. Weiss at Chicago. It is a ring about 1.7 microns in circumference. The original light-scattering data indicated that this DNA had a molecular weight of between 1.6 and 1.7 million, and it therefore was calculated to contain somewhere between 5400 and 5500 nucleotides. We now know from the sequence data that it is actually 5386 nucleotides, exactly, which is certainly as good a confirmation as we might have hoped. The evidence initially that it was single-stranded, of course, came from a variety of indirect means. One can now in fact do this sort of thing with an electron microscope where one can differentiate a partially double-stranded region from single-stranded regions (Figure 10). Another picture (Figure 11) of ØX DNA shows the cistron-A protein, to which I will refer in a moment, attached at a particular point.

Now, after these basic physical parameters were determined, in order to get at the viral functions, we borrowed the technique to which Max referred earlier, that of conditional lethal mutants, and set out to use that technique to

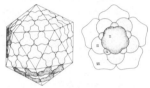

FIGURE 7 — *Schematic reconstruction of structure of ØX174 virus.*

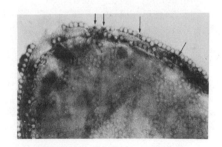

FIGURE 8 — *ØX174 particles attached to the cell surface.*

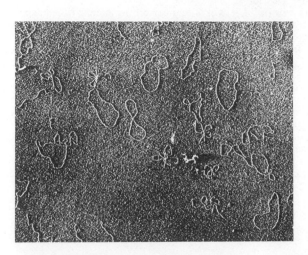

FIGURE 9 — *Single-stranded DNA rings from bacteriophage ØX174 (courtesy of Prof. Weiss).*

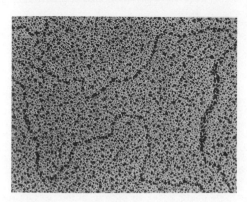

FIGURE 10 — *ØX174 DNA in which a portion of the ring is double-stranded.*

FIGURE 11 — *Cistron A protein attached to ØX DNA strand (Eisenberg et al., Proc. Nat. Acad. Sci. USA 74, 1977).*

determine the genetic structure of the virus. This resulted, over a considerable period of time, with the help of many students and postdocs in the establishment of a genetic map (Figure 12). At the time when we had eight genes, we attempted to introduce some logical order into the genetic structure and labeled them A-H; then, of course, another one came along to confuse that. But by means of these conditional lethal mutants, ambers, temperature-sensitives, ochres, and opals, we were able to establish the existence of nine complementation groups.

By combining the genetic data with studies which asked which of the events of the normal process of infection did not happen in the case of infection with a conditional lethal, we were able to assign the viral functions to the several complementation groups. The virion proteins were associated with genes J, F, G, and H. As I mentioned, F was the capsid protein, G and H the spike proteins, and J the small internal protein.

It was easy to demonstrate that gene A played some essential role in the replication of the double-stranded intermediate in DNA replication. Gene E was necessary for lysis, genes B, C, and D played roles of some sort in the production of progeny single strands.

I might say that at this point it was also possible by means of appropriate labeling experiments to begin to associate certain proteins made during infection with these particular cistrons. We then found ourselves in a rather awkward situation in that if one added up the molecular weights of the various proteins, it was touch and go whether one could accommodate those molecular weights, translated back into nucleotides, within the number of nucleotides believed to be present within the genome. This problem further was complicated by studies particularly by the group in Holland, in which by genetic mapping they demonstrated rather conclusively that the gene B appeared to have parts of gene A on both sides of it. It seemed to be nested within gene A. The resolution to these kinds of questions came from the determination of the nucleotide sequence, principally in Fred Sanger's lab in England.

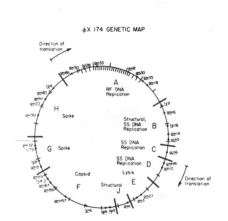

Gene	ØX174 (amino acids)	G4 (amino acids)
A	513	554
B	120	120
C	86	84
K	56	56
D	152	152
E	91	96
J	38	25
F	427	427
G	175	177
H	328	337

FIGURE 12 — Genetic map of bacteriophage ØX174.

FIGURE 13 — Sizes (number of amino acid residues) of proteins coded by bacteriophage ØX174 (Godson et al., in Denhardt et al., 1978).

Based on that sequence, to which I will refer in a few moments, we are able to determine of course the exact number of amino acids in each of the several proteins (Figure 13). You will note the largest is cistron-A protein, the next is F, and you will also note another protein K to which I will refer in a few moments, which was not picked up from the genetic analysis but by another means. G-4 is a virus similar to ØX which has been studied in Nigel Godson's lab with a similar range of protein sizes. The nucleic acid sequence (Figure 14) is shown here as determined in Sanger's laboratory. In order to make this interpretable, I have extracted portions of it and what I want to do now, in effect, is to walk through the sequence, pointing out features of interest.

FIGURE 14 — *The nucleotide sequence of bacteriophage ØX174 as determined in the laboratory of Prof. Sanger (Fiddes, Sci. Am. 237, Dec. Copyright.* © *1977 by Scientific American, Inc. All rights reserved).*

The DNA, of course, is a ring so where one starts is essentially arbitrary; but we will, for historical reasons, start just before the beginning of cistron A (Figure 15). What is presented here is a portion of the DNA beginning with the end of cistron H, terminating in a TAA. Then there is a gap, just for reasons of

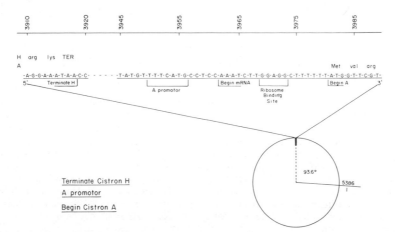

FIGURE 15 — *Details of ØX DNA sequence: termination of cistron H, A-promoter sequence, beginning of A-messenger RNA, A cistron ribosome-binding site, and beginning sequence for cistron-A protein.*

49

portrayal. Then one comes to the A-promoter and the sequence which begins the A-messenger RNA that had previously been determined by Lloyd Smith and Karl Grohmann. Next comes the ribosome binding site for the A-cistron protein which begins here with methionine, valine, arginine.

In Figure 16 we have moved on through part of the A-protein sequence, and one comes to that portion of cistron A which is the origin of viral strand synthesis. After the viral single strand enters the cell, it is converted to a double-stranded DNA ring by a process that I will describe. It is then transcribed and the product of cistron A is in fact a specific endonuclease that cuts the viral strand at a particular site, demonstrated here, which is, curiously enough, actually within cistron A itself. Presumably the cistron-A protein recognizes some portion of the sequence here and nicks at just that point, which then provides a site for the origin of subsequent viral strand synthesis. It is of some interest that that particular bond is in the midst of an AT-rich region and is surrounded by two GC-rich regions. There is a generally similar, although not identical, region in the G-4 virus As you will note, this sequence is right in the middle of the region which codes for the cistron-A protein--just going on in its normal way. In fact, Figure 16 compares the two regions in ØX and G-4 and in both cases there are AT-rich and GC-rich regions, although the sequences are by no means identical.

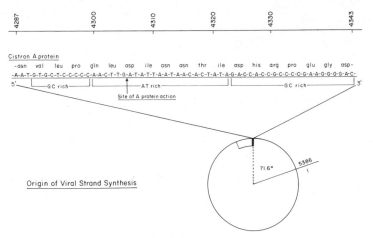

FIGURE 16 — *Details of ØX DNA sequence: origin of viral strand DNA synthesis (within cistron A region).*

Moving on in ØX DNA, now having gone through more of the ring, one comes (Figure 17), still reading the A-protein, to another promoter so as to begin reading another RNA message; again the initial sequence of this RNA had previously been determined. Then there is a ribosome binding site and then the beginning of the B-protein. Note that the virus is using the same nucleotide sequence as used for the A-protein but reading it in a different phase. And so when one comes to this ATG, which is part of a glutamic, asparagine sequence in the A-protein, in another phase, the ATG is read to be methionine, followed by glutamic, glutamine and so on, to start the B-protein. Thus one is reading this portion of the DNA in two phases, one for cistron-A protein, and one for cistron-B protein.

The DNA continues to be read in both phases, to produce both cistron-A and cistron-B proteins. Then one comes (Figure 18) to the end of cistron B, a TGA to terminate that, and one comes, actually in the <u>third</u> phase, to the beginning, with just a slight overlap, of the cistron-K protein; there is, in this phase, another ATG to begin the reading methionine, arginine, etc., and there is a short ribosome binding site just prior to that ATG.

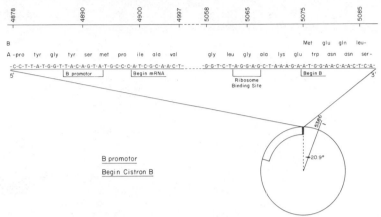

FIGURE 17 — *Details of ØX DNA sequence: B-promoter sequence, beginning of B-messenger RNA, ribosome-biding site, and beginning sequence for cistron-B protein (cistron A continues).*

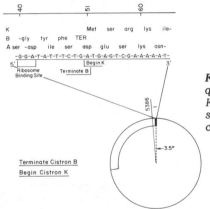

FIGURE 18 — *Details of ØX DNA sequence: ribosome-binding site for cistron K, termination of cistron B, and beginning sequence for cistron-K protein (cistron A continues).*

The cistron-K protein was not picked up genetically. It has been picked up as a protein in G-4 infection, not as yet, I believe, in ØX infection, although the sequence is so similar as to cause one to believe it must be present. Its function is unknown.

So now the DNA continues to be read, in the original phase, to specify the cistron-A protein, while in the third possible phase, it is specifying the cistron-K protein.

Continuing (Figure 19), one comes finally to the end of cistron A and one comes in the second phase to an ATG, which is the beginning of the C cistron and there is a slight overlap there. There is a ribosome binding site for the beginning of C. Meanwhile, in its phase, cistron K continues to be read; that is, K overlaps between A and C. Finally, then, one comes now (Figure 20) (in the middle of the C-protein) to the end of the K cistron, and for the first time in a good many nucleotides the DNA is read in only one phase.

Then (while reading cistron C) one comes (Figure 21) to the D-promoter, the beginning of the third messenger RNA, a sequence that had previously been determined. After the ribosome binding site for cistron-D protein, the termination of cistron C and the beginning of cistron D are again overlapped. Note that in practically every case the termination of one cistron and the beginning of the next either partially or wholly overlap. There is virtually no space in between the successive cistrons.

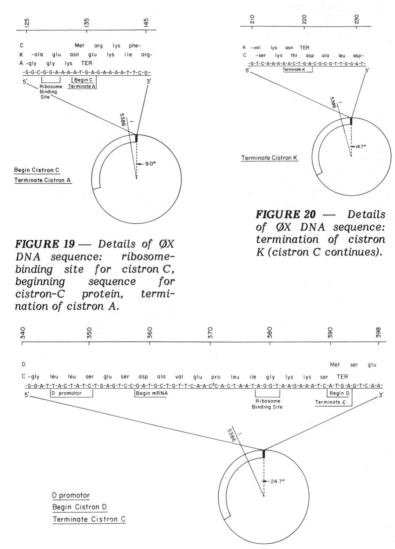

FIGURE 19 — Details of ØX DNA sequence: ribosome-binding site for cistron C, beginning sequence for cistron-C protein, termination of cistron A.

FIGURE 20 — Details of ØX DNA sequence: termination of cistron K (cistron C continues).

FIGURE 21 — Details of ØX DNA sequence: D-promoter sequence, beginning of D-messenger RNA, ribosome-binding site for cistron D, beginning sequence for cistron-D protein, termination of cistron C.

Then within cistron D one comes (Figure 22) to an ATG which is the beginning of cistron E, clearly in a different phase from that in which D is read (there is also within cistron D a ribosome binding site for cistron E). At the well known amber-3 mutation site, a mutation from G to A introduces a termination for E and therefore blocks formation of the cistron-E protein, producing a lysis defect; this mutation does not change the reading for the D-protein because of a codon redundancy in the D reading frame.

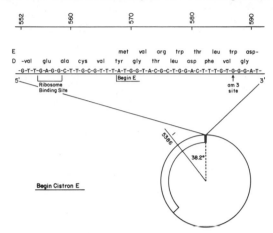

FIGURE 22 — Details of ØX DNA sequence: ribosome-binding site for cistron E, beginning sequence for cistron-E protein, site of am3 mutation (cistron D continues).

Then one comes (Figure 23) to the end of cistron E, a TGA, followed by the end of cistron D, and the beginning of cistron J, once again overlapping the termination codon of D. J is a short protein, as I have indicated--only 38 amino acids.

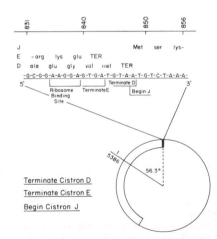

FIGURE 23 — Details of ØX DNA sequence: ribosome-binding site for cistron J, termination of cistron E, termination of cistron D, beginning sequence for cistron-J protein.

Continuing, one comes (Figure 24) to the end of cistron J and then for the first time one has a stretch of nucleotides that, at least seemingly, are not used for anything, before one comes to the ribosome binding site at the beginning of cistron F. F is a very large protein. At the end of F (Figure 25) there is, again,

53

an extended series of nucleotides which at least seemingly are not used before one comes to the beginning of G. The end of G is followed closely by the beginning of cistron H (Figure 26). One then runs through H, back to where we started on the first figure.

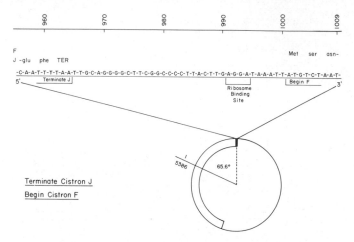

FIGURE 24 — *Details of ØX DNA sequence: termination of cistron J, ribosome-binding site for cistron F, beginning sequence for cistron-F protein.*

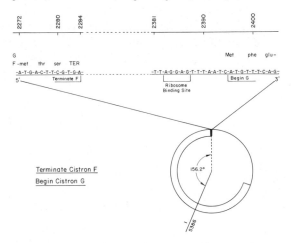

FIGURE 25 — *Details of ØX DNA sequence: termination of cistron F, ribosome-binding site for cistron G, beginning sequence for cistron-G protein.*

Now all of this clearly indicates that where there are sequences which <u>can</u> serve as ribosome binding sites followed by an ATG (in almost any of the possible phases) the cell seems to make use of such. This observation inspires one to look at the entire sequence to ask, "Well, are there other such sequences that may initiate proteins that we just may not have picked up for one reason or another?" And indeed (Figure 27), one can see there is a whole set of ATGs which could serve as the possible initiation codons for proteins of various lengths. Whether or not these are in fact used in the cell is still unknown. I have illustrated (Figure 28) how one of these might go; within cistron H there is an ATG (or another one further along) and there are possible ribosome binding sites that could be used. Thus, within the H-protein, one could start another protein at one or both of these two sites. Whether these are in fact used is not yet known. Each of these would give rise to a protein of some hundred or so amino acids before one comes to a termination in that phase.

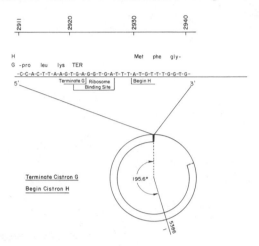

FIGURE 26 — *Details of ØX DNA sequence: termination of cistron G, ribosome-binding site for cistron H, beginning sequence for cistron-H protein.*

Clearly, the purpose of all these proteins is to ensure the reproduction of the virus and, of course, the production of the progeny virion. The sequence of events during infection has been worked out in considerable detail by a rather extended series of studies. As I have implied, the single strand of DNA after entering the cell is converted into a double-stranded DNA (Figure 29). It is then replicated as such to produce on the average some 15 or 20 double-stranded rings. These serve as templates for the production of messenger RNA. Then at a later stage of infection the double-stranded rings are used as templates for the production of, typically, one- or two-hundred single-stranded rings which are packaged into progeny virions essentially as rapidly as they are made. This sequence is subsequently followed by cell lysis.

Name	Position	Sequence	Length of protein
A1a	4186	G.G.C.G.T.T.[G.A.T.G.].T.T.C.G.A.T.A.A.T.C	25
A1b	4198	G.A.T.A.A.T.[G.G.].G.A.T.A.T.G.T.A.T.C	21
A2	4429	C.T.T.[A.A.C.G.A.].T.A.T.T.C.G.G.C.G.A.T.C	23
A3a	4621	C.G.A.T.T.A.[G.A.G.A.G.].C.G.T.T.T.T.A.T.C	78
A3b	4699	G.A.G.G.G.G.T.C.G.C.[G.A.A.G.].C.T.A.A.T.C	52
F1	1038	G.C.C.[G.A.].C.G.T.A.T.G.C.C.G.C.A.T.C	52
F2a	1272	A.T.G.[A.A.G.G.A.].T.G.G.T.G.T.T.A.A.T.C	38
F2b	1317	A.C.T.[G.G.A.].T.A.A.T.A.T.T.C.A.A.C.C.A.T.C	23
F3a	1449	A.C.C.[G.A.G.G.G.].C.T.A.A.C.C.C.C.T.A.A.T.C	34
F3b	1464	A.A.T.[G.A.A.].C.T.T.A.A.T.A.T.G.A.A.G.A.T.C	29
F4	1686	[G.G.A.G.G.G.].A.A.A.A.A.C.C.T.G.T.T.A.T.C	29
G1a	2543	G.A.T.A.G.T.T.T.G.A.C.[G.G.].T.A.A.T.C	62
G1b	2552	A.C.[G.G.G.].T.A.A.T.C.C.C.T.A.A.T.C	59
H1a	3076	A.C.T.G.T.[A.G.G.].C.A.T.C.G.G.G.T.A.A.T.C	100
H1b	3109	G.C.C.A.T.T.C.[A.A.G.G.G.].C.T.C.C.T.A.A.T.C	187'
H1c	3316	G.C.A.T.T.T.C.C.T.[G.A.G.].C.T.T.A.A.T.C	116
H1d	3340	G.A.G.C.G.T.G.C.T.[G.G.].G.C.T.G.A.T.C	108
H1e	3439	A.T.T.G.C.C.[G.A.G.].A.T.G.C.A.A.A.A.T.C	73
H1f	3508	A.C.G.A.A.A.A.G.A.C.C.[A.G.G.G.].A.T.A.T.C	52
H1g	3517	C.[A.G.G.G.].T.A.T.T.A.C.C.A.C.A.A.A.A.T.C	60
H2a	3784	G.C.A.[A.A.G.G.A.].T.A.A.T.T.C.G.T.A.A.T.C	48
H2b	3826	G.T.[G.G.].T.C.A.T.A.T.T.T.T.T.C.A.T.C	34
16S RNA		3' ₃ₒ[A.U.C.C.U.C.C.].C.U.A.G 3'	

FIGURE 27 — *Other possible protein initiation sites in ØX DNA (Godson et al., in Denhardt et al., 1978).*

Much of the enzymology of this reproduction has now been worked out in in vitro systems, principally in the laboratories of Arthur Kornberg at Stanford and Jerard Hurwitz at Albert Einstein. The conversion of the single-stranded DNA to the double-stranded ring form is performed, of course, by pre-existent enzymes since the single-stranded DNA cannot be transcribed. One of the most

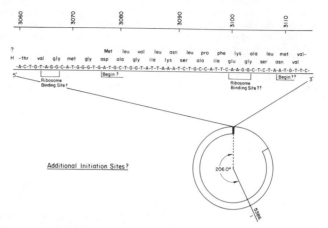

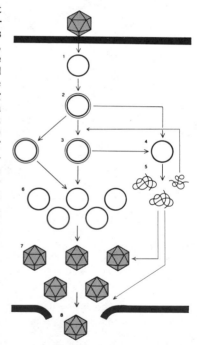

FIGURE 28 — *Two possible protein initiation sites within cistron H.*

remarkable aspects of this process is that the conversion from single- to double-stranded DNA involves quite distinct paths in ØX and in two closely related phages, G-4 or ST1, and still another path, in the rod-shaped single-stranded DNA phages, Fd or M13. All of the conversions from single to double strand involve what the Kornberg laboratory (I will use their nomenclature in order to keep matters simple; the Hurwitz laboratory has a different nomenclature) calls holoenzyme III, which is a complex of DNA polymerase III and three other proteins, elongation factors. The holoenzyme III acts to elongate the complementary strand once a primer has been formed. And the differences between the different viruses relate to the mode of formation of this primer. It is most complex in the case of ØX, involving all of the various proteins listed in Table I.

After the viral strand upon entering the cell becomes coated with the DNA unwinding protein (or DNA binding protein, as it is most commonly called), four other proteins--the host dna-B protein, which has an ATPase activity; the host dna-C protein; and the i and n proteins of Kornberg--are necessary in order to permit attachment of the primase (the host dna-G protein) to the ØX strand. Of these proteins, it is known

FIGURE 29 — *Sequence of events in ØX viral replication.*

that the dna-C protein and the i and n proteins act catalytically and are not needed once the dna-G protein has in fact been bound. Now the dna-B protein by itself can attach to the DNA coated with the unwinding protein and can migrate, processively, along the DNA, using ATP as a source of energy.

Presumably at certain positions in that migration, it is able in the presence of the other proteins to form a proper complex with the dna-G protein. And once that happens a primer can be formed using either ribonucleotides or deoxynucleotides; then if holoenzyme III is present, that primer can be and will be extended to form the complementary strand. In the absence of holoenzyme III, in vitro, formation of the primer will stop at a length of some fifteen or twenty nucleotides, the dna-B, dna-G complex will precess along the strand and a new primer may be initiated a few hundred nucleotides further along.

Protein	DNA binding[a]	NEM-sensitive	Gene locus	Molecular weight	Function
dnaC	+	+	dnaC	20,000	?[b]
dnaB	−	−	dnaB	250,000[c]	?
Protein i	±	−	?	40,000	?
Protein n	+	+	?	80,000	DNA binding
DNA polymerase III holoenzyme[d]	+	+	dnaE	330,000[e]	DNA synthesis
dnaG	+	−	dnaG	65,000[f]	RNA synthesis
DNA-unwinding protein	+	−	?	76,000[g]	DNA binding

TABLE I — *Proteins required for the formation of the complementary strand to ØX viral DNA.*

While all of this rather complicated apparatus is necessary for ØX, in the case of similar phages, G-4 or ST1, formation of the primer is much simplified. With these other phages, the dna-G protein (in the presence of DNA binding protein, ADP and deoxynucleotides) by itself is able to form a primer at a specific DNA site, in the absence of any of the other proteins required for ØX. The sequences of primer formed for G-4 and ST1 are very similar and presumably it is the absence of this particular appropriate sequence in ØX that prevents the use of the simpler mechanism and requires this much more complex process.

Interestingly, in the case of Fd or M13 the primer is formed not by dna-G protein but by the more conventional RNA polymerase at a specific site of the DNA when it is coated with DNA binding protein. And again this path is specific to these phages. RNA polymerase cannot be used with ØX or with G-4 to form a primer. Obviously, there are specificities of sequence that are not yet understood.

Having formed the double-stranded DNA, the events in the subsequent replication are shown in Figure 30. A supercoiled RFI is necessary, and supercoiling can be introduced into non-supercoiled RF by the action of the enzyme known as DNA gyrase in the presence of ATP. Once one has the supercoiled RF, this is nicked at the specific site I mentioned previously by the product of the cistron-A protein. The cistron-A protein then links covalently to the 5'-end after nicking. Subsequently, in the presence of the host rep protein, ATP, and DNA binding protein, the opened viral strand will begin progressively to unwind from the complementary strand. If only these components are present, as shown here on the right, this unwinding will proceed to completion, and one will end up with the closed minus strand and nicked viral strand with the cistron-A protein bound to the 5'-end. If, however, holoenzyme III is present and deoxynucleotide-triphosphates (and again the rep protein and the DNA binding protein), then as the viral strand is unwound, the 3'-end of the viral strand will be extended by the action of holoenzyme, using of course the complementary strand as template.

57

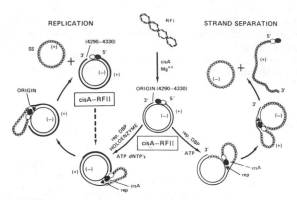

REPLICATION   RFI   STRAND SEPARATION

FIGURE 30 — *Role of the cistron-A protein in the replication of ØX DNA. In the absence of holoenzyme III, the result is strand separation. In the presence of holoenzyme III, the result is release of a viral single-strand DNA and regeneration of a double-stranded RF DNA (from S. Eisenberg, J. Griffith and A. Kornberg, Proceedings, National Academy of Sciences 74, 3198-3202, 1977).*

Furthermore, the cistron-A protein, which as you will recall is attached to the 5'-end, also attaches itself to the replicating complex and so remains looped back at the replicating fork. When the replicating fork has gone all the way around and past the origin, generating again that sequence which is susceptible to the cistron-A protein, the cistron-A protein then acts to cut the viral strand. In the process it links the original 5'-end to the 3'-end after cutting, thereby producing a closed single viral strand. The cistron-A protein at the same time attaches itself to the new 5'-end so that it can go around again. (It is very clever.)

In <u>vitro</u> this can be carried out to yield as many as 15 or 20 single strands from one single RF ring. In <u>vivo</u>, of course, other factors are present. And when this happens, when the viral single strand is produced in the early stages of infection it is converted back to a double-stranded ring by the action of the same cluster of proteins that I previously mentioned; or later in the infection, when the viral B, D, F, G, and H proteins are present, it is encased into a virion.

The details of this latter process, which is now accessible to <u>in vitro</u> experimentation, are still not fully understood. There is evidence of intermediates involving particularly the cistron-D protein, which is then subsequently lost in the production of the mature virion. The role of the viral D protein and the role of the viral B protein are still not fully understood. In all of this, the fact, however, that each of these several stages can now be carried out essentially independently in the test tube makes it a certainty that the details will become understood.

What then are the current problems? Well, the obvious one is to understand more of the details of the formation of virions. There are also from the point of view of enzymology many ill-understood aspects of what I have just described in fairly simple outline. The function of the viral cistron-C protein is as yet a mystery. Another aspect to this that I will just mention briefly has to do with the fact that there is a specific methyl group, a methylcytosine in ØX DNA. It is located elsewhere in the cistron-H protein. Its function is unknown and the enzyme that generates it is unknown. It is known that when ØX infects, a new cytosine methylase then appears. (Mutants can be used to infect <u>E. coli</u> B which does not itself have a cytosine methylase.) Whether this is one of the proteins or whether it is somehow an induced host protein is not understood.

Portions of the genome can be introduced (and have been introduced in Chambers' laboratory) into plasmids where they can function. Cistron G in particular has been introduced into a plasmid and it can provide cistron-G protein and this provides opportunities, of course, to do some interesting experiments.

Furthermore, since the entire sequence is known, it is possible to introduce mutations at particular regions anywhere one desires in the genome, and then to generate whole phage therefrom. Thus one can produce a wealth of mutations in any of the proteins, in order to see the subsequent effects; alternatively, mutations could be induced in any of the DNA binding sequences, the promoter sites, the ribosome binding sites, the initiation sites, etc.

This paper is, I think, a reasonable summary of the present status of understanding of the biology of this virus. Much of what has been learned from ØX research is, of course, unique to this type of virus. Some is of more general interest. Other single-stranded DNA viruses, plant, animal, and bacterial, are now known. Essentially they are one of the basic subsets of living creatures. Circular ring DNAs, of course, have been found in many and varied settings.

More broadly, I think we have come to appreciate the reproduction of a DNA as an intricate process so that in a sense if the virus supplies less of the machinery, then the virus must be so adapted as to make greater use of that available from the host. Curiously, as have been demonstrated with ØX, G-4, and M13, this adaptation or parasitism can follow varied paths, some much more complex than others, and yet all of apparently roughly comparable advantage.

In the reproductive process the virus itself must probably contribute some unique step so as to provide a distinct advantage to the selection of viral DNA for multiplication as compared to the other DNAs available. We have also in ØX an example, not unique and not yet wholly understood, of the coupling of DNA reproduction with its packaging, which is a process of evident economy although also complexity. And, of course, as is demonstrated in the sequence, one has a superb example of the economy and the efficiency that can be achieved by evolutionary pressures. The multiple readings of the same DNA sequence in different phase, driven presumably by the constraint upon total virion DNA content, markedly expand our perception of the potential that is latent in a given DNA sequence. Once again the ingenuity of nature has outstripped our pedestrian imagination.

The last conclusion I would draw from this retrospective has to do with the collectivity of research. We now know a lot about ØX, and this has come about obviously from the studies of a great many people in a large number of laboratories. It could not have come about in any other way. It has needed virologists, molecular biologists, enzymologists, electron microscopists, physical chemists. It's an impressive example, I think, of the power of modern biology when it is focused on a limited set of objectives.

Selected Readings

Benbow, R. M., C. A. Hutchison III, J. D. Fabricant and R. L. Sinsheimer. Genetic map of bacteriophage ØX174. Journal of Virology 7:549-558 (1971).

Dressler, D., D. Hourcade, K. Koths and J. Sims. The DNA replication cycle of the isometric phages. In: The Single-Stranded DNA Phages. D. T. Denhardt, D. Dressler and D. S. Ray, eds. Cold Spring Harbor Laboratory: Cold Spring Harbor, New York, pp. 187-214 (1978).

59

Godson, G. N., J. C. Fiddes, B. G. Barrell and F. Sanger. Comparative DNA sequence analysis of the G4 and ØX174 genomes. In: The Single-Stranded DNA Phages. D. T. Denhardt, D. Dressler and D. S. Ray, eds. Cold Spring Harbor Laboratory: Cold Spring Harbor, New York, pp. 51-86 (1978).

McMacken R. and A. Kornberg. A multienzyme system for priming the replication of ØX174 viral DNA. Journal of Biological Chemistry 253:3313-3319 (1978).

Sanger, F., G. M. Air, B. G. Barrell, N. L. Brown, A. R. Carlson, J. C. Fiddes, C. A. Hutchison III, P. M. Slocombe, M. Smith, J. Droulin, T. Friedman and A. J. H. Smith. The nucleotide sequence of the DNA of ØX174 cs70 and the amino acid sequences of the proteins for which it codes. In: The Single-Stranded DNA Phages. D. T. Denhardt, D. Dressler and D. S. Ray, eds. Cold Spring Harbor Laboratory: Cold Spring Harbor, New York, pp. 655-669 (1978).

Sinsheimer, R. L. Purification and properties of bacteriophage ØX174. Journal of Microbiology 1:37-42 (1959).

THE RISE AND DECLINE OF T4 PHAGE BIOLOGY AT CALTECH:
A LATTER-DAY VIEW

William B. Wood, Chairman

Department of Molecular, Cellular, and Developmental Biology
University of Colorado, Boulder

I would like to begin with some early history. It will be from a different perspective, partly because I am still actively involved in phage research, and partly because I was only about six months old when the first one-step growth experiment was carried out here. It all began with Max Delbrück, as you know. I can tell you no more than you have heard already about those early years--the collaboration with Emory Ellis and the formation of the first phage group here at Caltech. During the succeeding period, bacteriophage T4 gave us, among other things, strong corroborative evidence that the genetic material was DNA, with help from A. D. Hershey; the first proof for the existence of messenger RNA with help from Sidney Brenner, Francois Jacob and Matt Meselson; and our first real understanding of the functional relationships between fine structure elements of a gene, the rII gene, with help from Seymour Benzer.

By the early 1960's the major goals of Max and his original group had been realized, at least in a general sense, with the general solution of what we might call the genetic problem--that is, how is the phage DNA replicated and how does it direct the synthesis of viral proteins? Of course a lot of details were still to be learned. But at about that time new work at Caltech by Bob Edgar and Dick Epstein, who were then postdocs in the phage group, started phage research in a new direction.

Although the general solution to the genetic problem had been achieved, there remained (and still does remain) what we might call the epigenetic problem--that is: How is the temporal and spatial organization of gene product function controlled to form the phage particle, or for that matter, to form any living organism? This problem is as basic to biology as the genetic problem. It is still so far from solution that most molecular biologists have not even defined it as interesting. Certainly in recent times it hasn't yet become the focus of research that the genetic problem was in the 1950's.

Edgar and Epstein pointed phage research toward the epigenetic problem with their discovery of conditionally lethal mutants, which for the first time allowed a systematic investigation of essential viral genes--genes that are required for DNA synthesis, assembly of the phage particle, and so on.

Their work made it possible to explore essential gene functions by combining genetics with biochemistry, electron microscopy, and other techniques. And these studies, a few years later, led directly into investigation of how the phage particle is assembled.

Before proceeding to that topic, let me introduce you to T4 as we understand it today. Among the phages, T4 is in many ways at the opposite end of the spectrum from ØX. ØX succeeds by efficiency, simplicity, economy, and quite strong reliance on host cell functions. T4 by contrast succeeds by versatility, complexity, relative self-sufficiency and a certain amount of brute force in dealing with the bacterial host.

Figure 1 is a picture of T4 taken several years ago by Robley Williams. You can see that in contrast to the classic icosahedral symmetry of ØX, T4 is a baroque contraption with a complicated contractile tail which allows it to transfer the DNA from its elongated head into the bacterial cell. T4 even has what appears to function as a rudimentary sensory apparatus in the form of slender whiskers that extend outward from the neck region. The whiskers have the capacity to hold the tail fibers in the upright position and thereby prevent infection of bacterial cells at low temperature or under adverse ionic conditions.

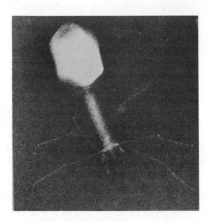

FIGURE 1

Like its morphology, T4's genetic repertoire and its strategy for host cell takeover are complicated. Figure 2 shows a taxonomy of T4 gene functions, to use a phrase that Bob Edgar liked to employ. About 140 T4 genes have now been genetically identified. Probably there are no more than about 20 not yet identified. The functions are known, or can be guessed at, for about 120 of the known genes. These can be divided into two major categories: those at the top of the chart that function in cell metabolism and those at the bottom that control the assembly of phage particles.

Many of the metabolic genes function in the synthesis of the phage DNA. T4 modifies its DNA and sticks glucosyl groups onto it by way of overcoming the host restriction machinery, and so it must make enzymes that are not found in the bacterial host. These enzymes were important in early work proving that after infection, phage don't simply induce bacterial genes; they express new genes of their own. Remarkably, 60 of the 80 metabolic genes shown here are nonessential under standard laboratory conditions, suggesting that T4 carries them as insurance. Most of these genes are not required in any particular bacterial host, but they allow the phage to have a more extended host range and to increase the production of progeny by supplementing the host metabolic machinery.

The bottom half of the chart in Figure 2 shows the genes involved in particle assembly. Let us turn now to that topic.

The morphological complexity of the virion and the availability of Edgar and Epstein's conditionally lethal mutations turned out to be a fortunate combination for investigating the morphogenesis of the phage particle. The general problem of morphogenesis can be simply stated as follows: how does the linearly arranged information in a DNA sequence specify the three-dimensional structure of a biological object like T4? Such understanding as there was about this process until recently was confined to multimeric proteins and simple viruses like tobacco mosaic virus. These structures can form by self-assembly, which can be defined as an assembly process in which all the necessary energy and information are contributed by the assembling components

62

which comprise the final product. No tools are required; the parts alone can assemble spontaneously by themselves.

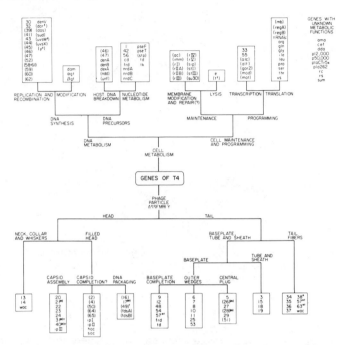

FIGURE 2 — From W. B. Wood and H. R. Revel, The genome of bacteriophage T4. *Bacteriological Reviews* **40**, 847-868 (1976).

The approach to studying assembly of these simple structures had been to dissociate them with mild chemical procedures and then to observe their reformation. There are two disadvantages to that approach. First, the reformation may occur by a pathway different from that of the original assembly process in the cell. Second, this approach will not work with anything much more complicated than tobacco mosaic virus. T4 cannot reassemble following dissociation. However, the conditionally lethal mutants of T4 allowed a different approach: an opportunity to genetically dissect the original process of assembly, from virgin proteins of a structure considerably more complicated than tobacco mosaic virus.

Figure 3 is a simplified genetic map of T4, emphasizing just the genes essential for the assembly process as defined by Edgar's and Epstein's work. The boxes on the outer circle contain schematic diagrams of the structures that are seen in electron micrographs of cells infected with mutants defective in a particular gene. From these pictures, one can infer the nature of the various gene functions. For example, when there is a mutation in gene 23 or one of the neighboring genes, the infected cells accumulate tails and tail fibers but not heads, so these genes are assumed to be head genes. By the same kind of reasoning, another group of genes controls the tail fiber; another group is essential for tail assembly, and so on. This work in the early 1960's already told us something about how the particle is put together. It seems to be made of at least three independent subassemblies--head, tail, and tail fibers--each of which can be built independently of the other two. If one of them is defective, the others still are made.

63

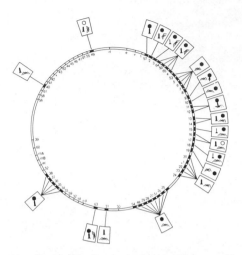

FIGURE 3 — *From W. B. Wood, Bacteriophage T4 assembly and the morphogenesis of subcellular structure. In* The Harvey Lectures, *Series 73, Academic Press: New York, pp. 203-223 (1979).*

This work also showed that a large number of genes is involved in the assembly process. The current estimate is about 50, which is more than the number of proteins in the phage particle. This discrepancy suggests that T4 morphogenesis may not be a simple self-assembly process, like that of tobacco mosaic virus reassembly, for example. Rather, the finding of accessory proteins necessary for assembly but not present in the finished virus implies departures from self-assembly, and suggests that T4 morphogenesis might involve mechanisms common to more complicated assembly processes in other cells.

To further understand the process, we needed more information than was provided by electron micrographs. The opportunity for getting some of this information came in 1965 with the discovery of in vitro complementation--that is, the finding by Bob Edgar and myself that the phage parts that accumulated in various mutants could be mixed together to form infectious virus in the test tube.

I remember the excitement of that time. We were doing experiments that probably neither Bob nor I would have done alone. The whole idea of making an extract of infected cells was very repugnant to Bob, who was a classical genetic purist, and I, trained primarily as a biochemist, was still intimidated by all the fancy mutants that were being used in the phage group. We talked about doing such experiments a year before I came to Caltech. During the first year I was here, we tried mixing purified tail fiberless particles, obtained from cells infected with a tail fiber–defective mutant, with an extract that contained tails and tail fibers, made from cells infected with a gene 23 (head-defective) mutant. Then, we took samples at various times from the reaction mixture and "plated" them to determine the level of infectious virus. I can remember coming in the next morning and being amazed and very excited to find that the plates were covered with plaques. It looked as though there had been a thousand-fold increase in the number of infectious virus during the first few minutes of the reaction!

We sat around all morning trying to figure out something that could have gone wrong or some trivial way to explain this result other than the one we hoped for: namely, that the tail fibers had attached to the phage particles in vitro to form infectious virus. Subsequent control experiments convinced us that tail fiber attachment was really occurring. Moreover, many other combinations of mutant-derived extracts could be mixed together with production of infectious virus. We were observing in vitro complementation of assembly mutants!

These results were important, aside from just their shock value at the time. They showed that the structures accumulated by these mutants probably were intermediates in normal assembly. Moreover, they allowed us to clarify some of the steps that were unclear from electron microscopy, for example, those defined by mutants that seemed to make all the parts but did not put them together. For the first time, we had a way to ask whether the parts seen in electron micrographs could actually function in assembly. Often we found parts that appeared normal but were nonfunctional in assembly, and therefore presumably incomplete. In general, in vitro complementation opened up the whole problem to a direct biochemical approach, in that we could study assembly in cell-free extracts and, like enzymologists, try to purify the components involved and understand the mechanisms of the assembly reactions.

By 1968 we had been able to formulate the pathway of T4 assembly shown below. It was based on a combination of more detailed electron microscopy, mostly by Jon King who was a graduate student of Bob Edgar's, plus many more in vitro complementation experiments. The pathway in Figure 4 shows the sequence of gene-controlled steps in the assembly process as we understood it then. The numbers are gene numbers. The solid arrows indicate steps that could be carried out in extracts; the dashed arrows indicate steps that at that time were not accessible to in vitro complementation experiments. The nature and the order of these latter steps were inferred from electron microscopy alone.

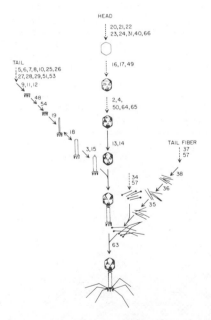

FIGURE 4 — From W. B. Wood, Bacteriophage T4 assembly and the morphogenesis of subcellular structure. In The Harvey Lectures, Series 73, Academic Press: New York, pp. 203-223 (1979).

The experimental work leading to the pathway confirmed the notion that the phage was made as three independent subassemblies which were then put together. It showed that for the moment we could study only the later steps in the process by in vitro complementation. Most interestingly, it showed that there is a strict order in the sequence of assembly steps. We knew that this sequence could not reflect a temporal sequence of gene expression. All the proteins in the particle are made simultaneously during the latter part of the phage infectious cycle. Therefore, the order that we observed had to be imposed at the gene product level, in the interaction of the proteins with each other.

At this point there were several changes in the makeup of the group here at Caltech. Bob Edgar left to work on a more complicated morphogenetic process: that of a college at U.C. Santa Cruz. Jon King left for the MRC Laboratory in Cambridge, England, to take a postdoctoral position there, and then subsequently joined the faculty at MIT. A significant technical development occurred: the development of polyacrylamide gel electrophoresis allowed us to identify the various phage gene products as bands on gels. These gave us assays for the gene products and allowed us to isolate individual parts of phage and ask which proteins they contained.

My group stayed here and continued work primarily on the tail fiber branch of the pathway. Jon King, particularly after his arrival at MIT, focused his attention on the tail. Uli Laemmli, who was here briefly as a postdoctoral fellow, and other workers trained in Edouard Kellenberger's laboratory in Geneva, focused their attention on the head. We had a sort of gentlemen's agreement that each of us would stick to these different parts of the pathway, and it worked fairly well. We didn't step on each other's toes or waste time competing with each other, and fortunately it turned out that each branch of the pathway posed different questions and showed quite different features, all of interest. I would like to go through each branch briefly in turn.

I will start with the tail, which was Jon King's main concern. Before he left here, Jon already had learned something about the final steps in its assembly. After the baseplate is formed, the next step is formation of the tube and then the sheath around the tube. Then a connector is added, and then the tail can attach to the head. The major remaining problem was assembly of the baseplate, a complicated machine whose construction was known to involve the action of at least 15 different gene products. In the early 1970's, Jon and his coworkers accomplished the remarkable feat of dissecting this pathway all the way back to its beginnings, as shown in Figure 5 (courtesy of Peter Berget and Jonathan King).

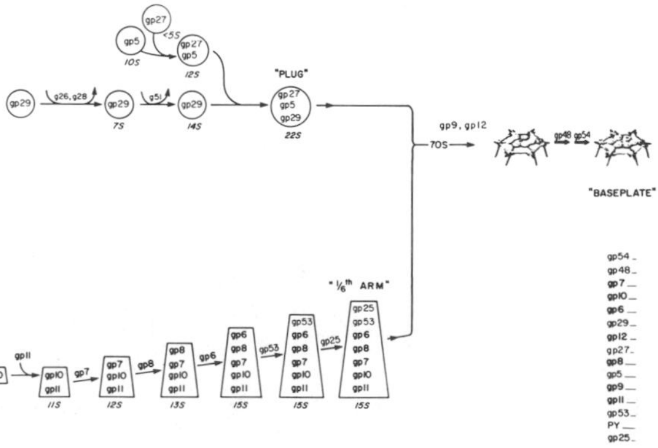

FIGURE 5

Their work involved finding in vitro complementation conditions for each of these steps, isolating each intermediate structure, and analyzing it to determine which gene products it contained. The pathway is branched; one series of steps leads to "plug" formation and another leads to the formation of a "wedge". Six wedges then assemble themselves around one of the plugs to form the hexagonal baseplate structure.

The principal take-home lesson from baseplate assembly is that this intricate machine is put together almost entirely by a complex self-assembly process. That is, with only two or three possible exceptions, each step of the pathway involves addition of another protein to the structure. Moreover, these steps occur only in the order shown. The baseplate is a very stable structure; it can be completely dissociated only by boiling in the presence of a strong denaturant. Clearly, there are strong interactions between the proteins in the finished product, and yet no one association step in its assembly will occur until the preceding one has occurred. For example, if the gene 10 product (gp10) is missing, all these proteins remain unassembled and free in the cell cytoplasm. We will return later to how this ordering might be accomplished. A possible clue is provided by two exceptions to the rule. Late in an infection that is lacking one of the initial gene products, some aberrant structures do start to assemble. The sheath subunits begin to form a dead-end structure called polysheath, which takes them out of circulation for subsequent sheath assembly. Also, the wedges in the absence of plugs slowly form unstable hexamers.

Now let us turn to the head, which turned out to be more difficult. Mutants that are blocked in head assembly tend not to accumulate true intermediates, but rather aberrant structures that cannot serve as components for in vitro assembly. This feature made it impossible to push the in vitro complementation work back much farther than it was in 1968. Nevertheless, a considerable amount has been learned, some of it from other phages that turned out to be more amenable to in vitro complementation analysis of head assembly. Study of these phages has led to some insights into T4, because all phages seem to assemble their heads in more or less the same way.

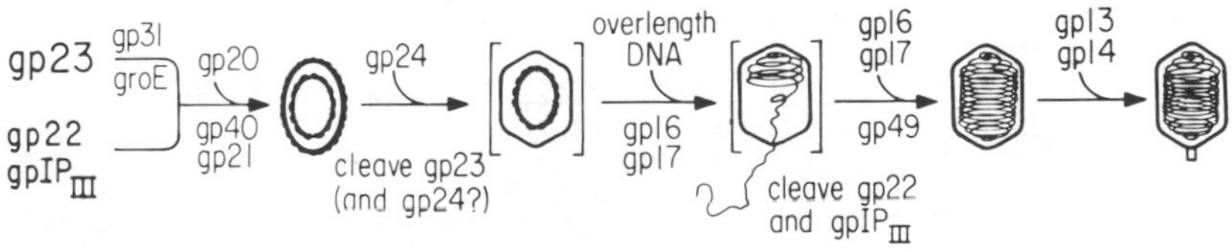

FIGURE 6

Figure 6 is a diagram showing a very superficial outline of how T4 does it. T4 and other phages begin, surprisingly, not by condensing their DNA into a coil around which they build a protein coat, but rather by making a protein shell and then introducing the DNA into it. The so-called prehead is assembled as a two-layered structure consisting of the outer shell and an inner protein core or scaffold which is required probably to make the head the right shape. We don't yet understand the causal relationships between the subsequent events. First, the core disappears. In T4 it is degraded into small peptides; in other phage the core proteins exit from the prehead intact and can be reutilized for further prehead assembly. The DNA enters the head, and the outer shell undergoes a "lattice transition", that is, a conformational change of all the subunits that increases the volume of the head. In some phages, including T4, these events are accompanied by proteolytic cleavages of several of the head subunits, in a fashion analogous to zymogen-to-enzyme conversions.

Somehow these machinations must solve the essential thermodynamic problem of DNA packaging: the DNA must enter and coil inside the head with a net free energy decrease, and yet the head of the finished virus must be a metastable structure from which the DNA can exit spontaneously into the host cell. Packaging is probably an energy-driven process, but its mechanism is not yet understood. Understanding it is particularly important because it is a process common to all viruses. Moreover, it probably occurs only in virally infected cells and not in normal cells, so that it is likely to be a good point for pharmacological intervention in viral infections. Several laboratories including our own are actively working on this problem.

Now let me turn to the tail fiber branch of the pathway. The tail fiber is relatively simple and its assembly pathway is the simplest part of the overall scheme. However, it presents a new challenge: although the finished product, shown in Figure 7, contains only four structural proteins (gp34, gp35, gp36, and gp37), eight phage genes are essential for building the structure and attaching it to the phage. The process appears to involve no breakage or formation of covalent bonds that would require catalysis by enzymes. What could be the

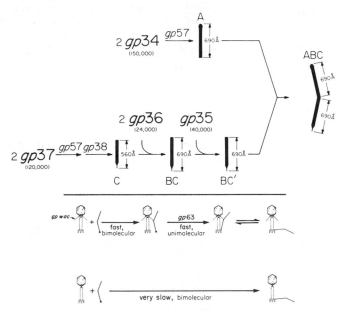

FIGURE 7 — From W. B. Wood, Bacteriophage T4 assembly and the morphogenesis of subcellular structure. In The Harvey Lectures, Series 73, Academic Press: New York, pp. 203-223 (1979).

functions of the four accessory gene products required in addition to the structural proteins? The pathway of tail fiber assembly shown in the upper half of the figure is branched; the proximal half of the structure (A) is made independently of the distal half (BC). Two of the four accessory proteins are required for assembly. The product of gene 57 is required for each of the two initial steps, which involve the dimerization of large polypeptides into a rod-shaped tail fiber precursor. For the distal half fiber, this step also requires a second accessory protein, the product of gene 38. After the other structural proteins are added and the two halves are joined, the whole fiber is attached to the bacteriophage particle as shown in the lower half of the figure. This is the last step of the assembly process in the cell. It involves two more accessory proteins. One is the whiskers, gpwac (whisker antigen control) and the other is the product of gene 63.

We believe that these two proteins convert what would otherwise be a very slow bimolecular reaction to a fast bimolecular reaction, the reaction of the tail fibers with the cloud of whiskers around the phage (there are six whiskers) plus a fast unimolecular reaction which is promoted by gp63. We still don't understand how gp63 acts, but it appears that this protein and gp38 somehow promote noncovalent association of structural components, presumably by transiently interacting with them. We shall return below to how this might work.

At least two important general insights have emerged so far from research on phage assembly. One is the realization of the importance of kinetic controls in the assembly process, and the other is the recognition of accessory protein functions. The early attempts in the 1950's and early 1960's to understand the structure and assembly of small icosahedral viruses by Crick and Watson and then by Casper and Klug assumed that in such simple structures the free energy of subunit association favored one form, namely the correct one, over all others, so that one could explain assembly in terms of thermodynamic equilibria alone. The resulting concept of self-assembly largely ignored pathways of subunit association.

The recent work on more complex viruses that I have outlined for you certainly has corroborated the importance of self-assembly, but it also has shown us at least two features that cannot be explained on the basis of only thermodynamic equilibria. The first is the rigidly controlled temporal sequence of assembly steps, most of which can occur only in one particular order. In many cases, maintenance of this order probably is important for the assembly process. For example, sheath proteins must not be allowed to assemble with each other until the baseplate structure is ready to receive them. The second feature derives from the fact that these phages contain a number of proteins that are capable of associating with each other in more than one way, to give bonding patterns that are not strictly equivalent but only "quasi-equivalent." Casper and Klug showed that quasi-equivalent bonding is essential to form, for example, a large icosahedral shell. Such proteins necessarily are capable of polymorphic assembly, that is, they can assemble in more than one way to form alternative structures of roughly equal thermodynamic stability. In T4 the head proteins can form aberrant "polyheads" as well as true heads, and the sheath protein can assemble into polysheath as well as true sheath. Somehow, the phage is able to direct subunits into correct structures rather than aberrant ones, at least most of the time.

These two features can be explained by controls on the relative rates of individual assembly steps. Intriguingly, many of these controls in T4 seem to be built into the self-assembling structure proteins themselves. So far we can only speculate about the control mechanism, but a likely hypothesis would be as follows. Each association requires conformational changes in the interacting proteins in order to occur. These changes require energy, and therefore constitute an energy barrier to the association reaction, called an activation energy in chemical terms. As the assembly process proceeds, the binding of each subunit alters its conformation, thereby lowering the activation energy of the next step. Now this isn't a new idea; it is essentially the phenomenon of cooperativity that is familiar in the formation of repeating structures such as the tobacco mosaic virus helix. However, T4 illustrates a novel form of this phenomenon that we might call heterocooperativity: different kinds of protein subunits affect each other's association in such a way as to create an assembly pathway, in which each step increases the rate of the next one relative to all other possible associations. We can make a similar argument based on relative rates for how subunits may be directed into only one of several energetically

69

equivalent polymorphic structures.

The other general insight that has come out of this research is the recognition of nonstructural accessory proteins, which represent departures from self-assembly. They appear to be of at least three kinds so far. There are scaffolding proteins that participate in head assembly; there are real enzymes, such as the proteases that catalyze cleavages of subunits in head assembly; and then there is the intriguing third kind encountered in the tail fiber pathway. These latter proteins may represent a novel class of biological catalysts which promote noncovalent association of proteins. By analogy to enzymes, we speculate that they might act by transiently binding to and stabilizing an otherwise unfavorable subunit conformation that represents the transition state of an assembly reaction. Figure 8 is a fanciful diagram of how such a protein, the gene 38 product in the tail fiber pathway, might act by stabilizing an initial nucleation step in the association of the two monomers that make up the rod of the distal half fiber. Similar proteins could be more widespread than we now realize in the assembly of muscle, connective tissue, and subcellular organelles.

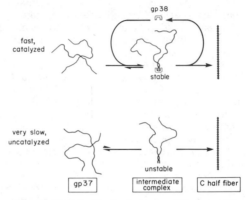

FIGURE 8 — *From W. B. Wood, Bacteriophage T4 assembly and the morphogenesis of subcellular structure. In* The Harvey Lectures, *Series 73, Academic Press: New York, pp. 203-223 (1979).*

That is all the science I want to present. I have told you about the rise of T4 phage biology at Caltech. But at the moment, phage research, at least for its own sake, is no longer being done in the Biology Division. What can account for the decline? Pinpointing the cause is difficult, but after looking at the evidence, I believe that something may be present in the halls of the Biology building here that induces the transformation of phage biologists into administrators. Sinsheimer and Edgar succumbed completely to this disease, and Wood is showing signs of the same pathology.

Whether there should be such a decline in prokaryotic research as there has been recently at Caltech and elsewhere is a general question too involved to go into here. Rather, let me in closing just mention an example that I think is instructive, and in which I was peripherally involved. Before coming to Caltech in 1965, I spent a year and a half as a postdoctoral fellow in the laboratory of a young Swiss biophysicist named Werner Arber. He was studying a remarkable process called DNA restriction which, he showed, allows bacteria to distinguish between their own DNA and foreign DNA, and to degrade the latter if it enters the cell. It was not apparent at that time that this work would be relevant to anything. As far as anyone knew, DNA restriction did not occur in eukaryotic cells, only in certain bacteria. It was an intriguing phenomenon that required an explanation; the consequences of understanding it were difficult to foresee. As many of you know, that work eventually led directly to the discovery of restriction enzymes which cut DNA at specific sequences. Last fall Arber was awarded a share of the Nobel Prize for Physiology and Medicine to recognize his part in developing what is now the most powerful and potentially valuable technology in modern biology. I think the lesson this should teach us is simply that wherever there is a particularly intriguing phenomenon in biology that we don't understand and that seems to demand an explanation, we shall probably be repaid for committing some time and effort and resources to investigating it. There are certainly many such phenomena still to be found among the prokaryotes and their viruses.

Selected Readings

Edgar, R. S. and I. Lielausis. Journal of Molecular Biology 32:263–276 (1968).
Edgar, R. S. and W. B. Wood. Proceedings of the National Academy of Sciences, Washington, 55:498–505 (1966).
Kikuchi, Y. and J. King. Journal of Molecular Biology 99:645–672 (1975).
Wood, W. B. Bacteriophage T4 assembly and the morphogenesis of subcellular structure. In: The Harvey Lectures, Series 73. Academic Press: New York, pp. 203–223 (1978).
Wood, W. B. and J. King. In: Comprehensive Virology. H. Fraenkel-Conrat and R. R. Wagner, eds. Vol. 13, in press (1978).
Wood, W. B. and H. R. Revel. Bacteriological Reviews 40:847–868 (1976).

THE LAST OF THE T PHAGES

F. William Studier

Department of Biology
Brookhaven National Laboratory, Long Island

In the summers of the early 1940's, a group of phage workers, including Max Delbrück, would meet at Cold Spring Harbor to do research. In order that the results of their individual efforts might easily be compared, they decided to concentrate on a set of phages that infect the B strain of Escherichia coli. Seven different phages were chosen, and were referred to as T1-T7, for types 1-7. Much of the early work on these phages concentrated on the closely related T2, T4 and T6, and many important discoveries about bacteriophages were first made using these phages. Bill Wood has just summarized some of the fascinating work on T4. T7, the seventh and last of the T phages, is the subject of my talk today.

The origin of T7 is an interesting story, which was recently related in a letter from Max:

> "T7 and δ were both isolated from a 'polyvalent' (mixed) phage supplied by an MD in New York. T7 was isolated by Demerec and δ, independently, by me. I then found out (1) that the alleged polyvalent mixture contained no other phage besides δ, i.e., no phage attacking B/δ; (2) that T7 and δ were serologically indistinguishable. When I told this to Demerec he suggested that we destroy one of these isolates and give out only the other. I do not recall whether this proposal was carried out..

> "A comical side note to this early history is the fact that I visited the supplier after I had found out that his mix contained only one phage that I could extract. I asked him how he made this polyvalent phage and he asserted that he grew the 40-odd phages separately and then mixed them. Afterwards I succeeded in talking to his technician separately and when I asked her how she prepared the polyvalent phage she said she had mixed the 40 phages and then grew them together to prepare the lysate. For all I know she may have subcultured the mixture through many subcultures!"

This story contains the germ of what I want to talk about, namely, genetic divergence of laboratory strains of T7, and the possible use of T7 to study evolutionary processes both in the laboratory and in nature.

In the time since it was named, T7 has become widely distributed among scientific laboratories around the world. As early as 1962, Davison and Freifelder reported that strains of T7 they received from Luria or Meselson, while identical in many respects, differed in electrophoretic mobility and in buoyant density in CsCl solution. Other T7 strains were also reported to fall into two classes, based on buoyant density. More recently, it was found that the two strains of T7 on which the most genetic work has been done, the strains used by Hausmann and myself, differ by at least a deletion and apparently some point mutations as well. The question naturally arises whether these differences among T7 strains arose through genetic divergence during or after the dispersal among different laboratories, or whether they might represent differences between the original T7 and δ strains.

A simple approach to answering this question became available with the discovery of restriction endonucleases that cut DNA molecules at specific sites, and the development of gel electrophoresis for resolving DNA fragments. The electrophoretic pattern of a restriction digest produces a rapid and reliable "fingerprint" of the DNA molecule. When comparing patterns given by closely related molecules, any deletions or insertions, or any point mutations that create or eliminate cleavage sites are evident at a glance. T7 is ideally suited to this kind of analysis, since phage stocks can be rapidly grown and purified, and the DNA can be released for cutting simply by brief heating of the phage particles followed by ethanol precipitation. It is possible to grow and analyze dozens of strains a week.

Having such a simple way to compare strains, it was easy to survey T7 strains from many different laboratories. The results of analyzing T7 strains from 19 different laboratories were a surprise to me: only eight appeared to be wild-type T7, nine were different derivatives of wild-type T7, and two were actually T3. The eight wild-type strains had descended from the early workers by different routes but were identical to each other in every test that was applied, including restriction analysis, plating behavior, and patterns of RNA and protein synthesis after infection. The nine derivatives of T7 were all different from each other and from the eight wild-type strains; most were pure deletion strains, but some were mixtures of different deletion mutants and wild-type, and some had detectable point mutations. These results seem to establish that the eight identical strains represent the original T7 wild-type, and that if T7 and δ were detectably different from each other, only one has had wide distribution. It is also clear that genetic divergence among different laboratory strains of T7 is fairly common. The two T3 strains presumably arose through mislabeling or contamination.

This result got me interested in the possibility of using restriction analysis to study the evolution of phages in the laboratory and in nature. A further test of restriction analysis was provided by a series of T7-related phages that had been thoroughly characterized by other means.

T7 turns out to be only one representative of a series of related phages that have been isolated around the world at different times. Table 1 lists six such phages, along with the places they were isolated and the dates of the first publications describing them. It was known from the early work on the T phages that T3 and T7 have the same morphology and are serologically related. Davis and Hyman, and Summers and his colleagues have explored the relationships among all of the six phages of Table 1 by heteroduplex mapping, two examples of which are shown in Figure 1.

Table 1. Some T7-related phages, their origins, and
the dates of the first publications describing them

T7	New York	1944
ØII	France	1947
H	California	1953
ØI	Italy	1961
W31	Japan	1964
T3	New York	1945

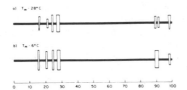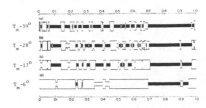

FIGURE 1 — *Heteroduplex analysis of the relationships between the DNAs of T7 and ØI (left) and between T7 and T3 (right). Apparent homology is indicated by the black bar, nonhomology by the open boxes, and position in the DNA molecule by the scale. Homology was measured at different temperatures below Tm, the closer to Tm the more stringent the requirement for homology. The T7/ØI analysis is from Hyman, Brunovskis and Summers, Virology 57, 189-206 (1974), copyright Academic Press, New York; the T7/T3 analysis is from Davis and Hyman, Journal of Molecular Biology 62, 287-301 (1971), copyright Academic Press, London.*

The DNAs of T7, ØII, H, ØI and W31 are quite closely related: heteroduplexes among these DNAs are well paired over most of the molecule, and only a few small regions of nonhomology are apparent (the T7/ØI heteroduplex of Figure 1 is typical). T3, on the other hand, is much less closely related to the other phages: heteroduplexes between T3 DNA and the other DNAs show partial homology over most of the molecule, with only perhaps one-fourth of the molecule appearing to be completely homologous (the T7/T3 heteroduplex of Figure 1 is typical). T3 also appears less related to the others by genetic tests: simultaneous infection by T3 and T7, for example, gives fairly strong mutual exclusion, most infected cells producing only one of the two phage types; and genetic recombination between the two phages is very poor. The phages of Table 1 are only a sampling of the T7-related phages so far isolated, but they are among the most thoroughly characterized.

Knowing how closely homologous the DNAs of T7, ØII, H, ØI and W31 appear by heteroduplex analysis, one might expect that the restriction patterns of these DNAs would also appear related. However, as shown in Figure 2, relationships are not apparent, except between variants of the same phage and between ØII and H. [The ease of comparing closely related strains can be seen by comparing the patterns for ØIIP, ØIIW and H: ØIIP is wild-type ØII; ØIIW is a deletion mutant of ØII that has fused two of the wild-type DNA fragments (about 2.5 and 1.3 kb in Figure 2) into a new band (about 2.2 kb); H is very similar to ØII, having a slightly longer deletion than ØIIW but in the same region of the DNA (the fusion fragment migrates atop a wild-type fragment about 2.1 kb).] Relationships among the other T7-related phages are not apparent from the restriction patterns, even knowing which restriction fragments of T7 DNA come from the long regions of good homology in the heteroduplexes. Thus, even with phages as closely related as these are, the restriction pattern of the DNA provides a characteristic fingerprint that can be used for simple and rapid identification; the restriction pattern is a very stringent criterion of identity.

Why should restriction patterns look so different when DNA molecules look so similar by heteroduplex analysis? If one takes a DNA of random nucleotide sequence, the six-base recognition sequence for any particular restriction endonulease would be expected to occur an average of once every 4096 bases. (T7 DNA is about 40,000 base pairs long, and would be expected to have 9-10 cleavage sites.) If the original base sequence is changed by random, single-base substitutions (≠ mutations), the probability of losing the original cleavage sites and creating new ones can be calculated. From such calculations, it can be estimated that random substitution of approximately 5% of the bases would eliminate about one-fourth of the original restriction sites and create an equal number of new ones. This would be more than enough change to make two restriction patterns appear to be unrelated, but would probably not be detected by routine heteroduplex analysis.

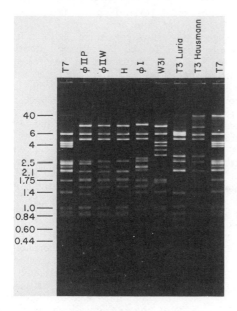

FIGURE 2 — Restriction patterns of DNAs from different T7-related phages. The DNAs were cut by the restriction endonuclease Hpal *and the products analyzed by electrophoresis in a 1.2% agarose gel. Approximate sizes are given to the left (in kb = thousands of base pairs). The picture is from Studier,* Virology **95**, 70-84 (1979).

This ability to distinguish closely related phage strains rapidly and easily by restriction analysis makes it possible to begin asking questions about evolution and ecology of phage populations, questions that would have been very difficult to answer before. For example, are the phages of Table 1 individual representatives of homogeneous populations of phages that were present in New York, California, France, Italy, and Japan at the time they were isolated, or do individual populations show a similar degree of diversity? If the populations are homogeneous, what are the relationships among different populations? What is the rate of genetic change in these phages? Are phages identical to those of Table 1 still to be found in nature today? The particular questions I would like to address now are, how does genetic divergence develop in T7-related phages, and is it possible to follow this divergence in the laboratory?

T7-related phages are ideal for approaching evolutionary questions not only because they are very easy to isolate, grow and identify, but also because they have a very short generation time. As shown first by Delbrück, new phages are released only 13 minutes after infection by T3 or T7 at 37°. Therefore, both in the laboratory and in nature, many generations can be produced in a very short time.

The different T7-related phages of Table 1 range from being almost identical to each other (ØII and H) to having very different nucleotide sequences in their genome (T3, relative to all the others). Yet these phages, and all other T7-related phages so far examined, share some rather constant features: the phage particles have identical morphology, even though their constituent proteins differ; the DNAs are all about the same size, and code for similar numbers of proteins; the genetic and functional organization appears identical;

the program of transcription is the same; and the patterns of protein synthesis are very similar, three groups of proteins being produced coordinately at different times after infection. The program of transcription is a particularly interesting feature of these phages: only a portion of the genome is transcribed by the host RNA polymerase, but one of the early phage genes specifies an entirely new RNA polymerase, and this new polymerase then transcribes phage DNA specifically.

Proteins having equivalent functions in T7-related phages usually have similar sizes, can often complement each other, and are specified by genes located at equivalent positions in the genome. However, some proteins have diverged considerably. The RNA polymerases of T3 and T7, although identical in function, have diverged to the extent that they transcribe the heterologous phage DNA very poorly. The recognition specificities of the polymerases and the transcription signals in the DNA have apparently diverged coordinately. Even greater differences are seen between the T3 and T7 proteins responsible for overcoming the DNA restriction system of the host. In both phages, the gene for this protein is at the extreme left end of the genetic map and is the first to be expressed after infection. However, the T3 protein acts by hydrolyzing S-adenosylmethionine, a cofactor required by the host nuclease, whereas the T7 protein appears to inactivate the host nuclease by direct binding. The two proteins behave differently on purification and may well be completely unrelated; and the region of the DNA that codes for them also appears unrelated in heteroduplexes. Yet, these two very different proteins perform the same function during infection and are expressed from genes located in the same relative position on the genome.

These kinds of analyses suggest that what is being conserved in the continuing evolution of the T7-related phages is a highly successful morphology and a highly successful genetic and functional organization. The sequence of nucleotides in the DNA, on the other hand, is apparently quite plastic and can vary widely. If this is so, one would predict that, as phages coevolve with their hosts in nature, phage strains should be generated that have exactly the same organization and structure as T7 but whose DNAs have absolutely no homology to T7 DNA.

Hausmann and his colleagues have examined just this possibility. They screened a series of phages that infect hosts that are different from but somewhat related to E. coli. Among a group of 20 phages having the same morphology as T7, they found several that have a T7-like pattern of protein synthesis after infection. Four of these induce a new RNA polymerase that specifically transcribes the DNA of its own phage but not heterologous DNAs, a characteristic feature of T7-related phages. Two of them induce an enzyme that hydrolyzes S-adenosylmethionine, an enzyme not previously known to be induced by any phage besides T3. The DNA of only one of these phages showed clear homology to T3 or T7 DNA by competition hybridization, and two showed indications of slight homology; but one showed no detectable homology (less than 5%) to either T3 or T7 DNA. These results seem to me to provide strong support for the idea that the morphological, genetic, and functional organization of the T7-like phages are strongly conserved but that the nucleotide sequence of the DNA is not.

If these ideas are correct, it might be possible to follow divergence of phage DNAs in the laboratory. Because of its rapid growth rate, many generations of T7 can be grown in a relatively short time. If the natural mutation frequency is high enough, or if it could be increased by growing the phage in the presence of a mutagen, changes might accumulate rapidly enough to be detected by restriction analysis, and ultimately it might be possible to produce phage strains in the laboratory that are as divergent as the phages of Table 1.

I have begun to look for divergence in the following way: starting with wild-type T7, a small lysate is grown; progeny from the first lysate are used to infect a second culture of E. coli B; progeny from the second lysate are used to infect a third culture; and so on for many cycles. Two independent series are being carried forward, one in normal culture medium and the other in the presence of the potent mutagen N-methyl-N'-nitro-N'-nitrosoguanidine. Each cycle of growth takes about one hour and produces about a 100-fold increase in the phage population in the absence of mutagen or about a 10-fold increase in the presence of mutagen. A small sample of each lysate is saved, so any divergence can be traced through intermediate states. By now, 45 cycles of growth have been completed, and the composition of the resulting phage populations has been analyzed by growing lysates from randomly isolated plaques and analyzing the restriction patterns of the individual phage DNAs.

Eleven individual plaques from the 45th cycle of growth in the absence of mutagen were analyzed. Interestingly, each of these strains carried a deletion, and these deletions were all in the same region of the DNA as the deletions found in the different laboratory strains of T7. Analyzing intermediate lysates in the series showed that the deletion mutants took over the population between the 10th and 15th cycles of growth. Obviously, such deletion mutants are able to outgrow wild-type T7 under these conditions, and this is the likely explanation for the frequent occurrence of such deletions in laboratory strains. Growth conditions in nature apparently do not favor the growth of such deletion mutants. Aside from the effects of the deletions, no changes in restriction patterns were found in the DNAs of the phages grown without mutagen.

Twenty-five individual plaques from the 45th cycle of growth in the presence of nitrosoguanidine were analyzed, and the restriction patterns of these DNAs, along with that of wild-type T7 DNA, are shown in Figure 3.

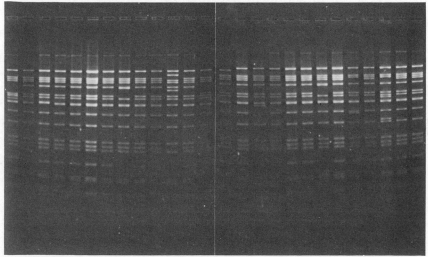

FIGURE 3 — *Restriction patterns of DNAs from T7 phages grown 45 cycles in the presence of mutagen. The pattern from wild-type T7 is at the left, and the patterns from 25 random single-plaque isolates are shown.*

Again, every strain has a deletion (two bands are missing from a cluster of six bands in the wild-type pattern, and the new band lies below a triple band of the wild-type pattern in all but one of the mutant strains). However, single-base mutations are also evident in these strains. Examination of the patterns in Figure 3 shows that 8 of the 25 phages have either lost or gained a cleavage site, and 1 of the 8 strains (the 11th pattern from the left) has lost two different cleavage sites. Assuming a random distribution of bases in T7 DNA and random base substitutions, it can be estimated that this population of DNA molecules has accumulated an average of around 100 base changes per molecule--that is, about 2-3 single-base changes per molecule per cycle of growth in the presence of the mutagen.

Each of the phages analyzed after growth in the presence of the mutagen had to be viable and had to have been able to grow in competition with the other phages in the population. In this sense, the 100 or so base changes accumulated per molecule could be called neutral mutations. Lethal mutations must also occur, and their frequency can be estimated by determining the specific infectivity of a population of phages produced after one cycle of growth in the presence of the mutagen. Such experiments indicate that growth in the presence of the mutagen produces an average of about one lethal mutation per DNA molecule per cycle of growth. Thus, the ratio of nonlethal to lethal mutations under these conditions seems to be on the order of 2:1 to 3:1, this in a DNA that is used almost entirely to code for proteins.

This series of experiments is still in its early stages, but it is already apparent that the rapid growth rate of T7, together with the ability of restriction analysis to detect changes in the total genome, will provide new opportunities for probing evolutionary processes. Continued serial growth and analysis of T7 in the absence of mutagen may make it possible to measure, or at least to put an upper limit on the natural mutation frequency of T7. The rate of accumulation of mutations in the presence of mutagens is high enough that it may be feasible to induce divergence at least as great as is found among the T7-related phages isolated from nature. Analysis of changes in individual proteins might also reveal patterns of evolution at the protein level. Particularly interesting would be the possibility of observing coevolution of the phage RNA polymerase and the transcription signals it recognizes in the phage DNA.

The idea of the early phage workers was to use phages as simple model systems in which to probe questions of general significance in biology. As we have heard in the previous talks, this approach has been spectacularly successful in molecular genetics and biochemistry. Evolution appears to be another area where phage work has much to offer.

Selected Readings

Davis, R. W. and R. W. Hyman. A study in evolution: the DNA base sequence homology between coliphages T7 and T3. Journal of Molecular Biology 62:287-301 (1971).

Hausmann, R. Bacteriophage T7 genetics. Current Topics in Microbiology and Immunology 75:77-110 (1976).

Hyman, R. W., I. Brunovskis and W. C. Summers. A biochemical comparison of the related bacteriophages T7, ØI, ØII, W31, H, and T3. Virology 57:189-206 (1974).

Studier, F. W. Relationships among different strains of T7 and among T7-related bacteriophages. Virology 95:70-84 (1979).

Studier, F. W. and N. R. Movva. SAMase gene of bacteriophage T3 is responsible for overcoming host restriction. Journal of Virology 19:136-145 (1976).

SESSION III: EVOLUTION, GENES, AND MOLECULES

Introduction

E. B. Lewis

Thomas Hunt Morgan Professor of Biology, Caltech

Formal genetics and molecular genetics are nicely represented in this session. Since the first speaker, George Beadle, needs no introduction to this group, I have time to recall here his immense contributions to the Biology Division. Almost all of us owe our existence, so to speak, to George. If Morgan was the grandfather of the Biology Division, Beadle was clearly its father.

Beadle saw perhaps more clearly than anyone else that the discipline of genetics had the power to probe the deepest secrets of biochemistry. His recent work is but another example of using genetics to attack one of the all-embracing and difficult problems of biology: the mechanism of evolution.

George wrote an introduction some years ago to a book that brought together the major publications of A. H. Sturtevant. I quote from that introduction: "I doubt if the excitement of scientific discovery can be appreciated in any real sense without personally experiencing it. Sturtevant experienced it abundantly." George Beadle also has not only experienced it abundantly, he never ceases to communicate that excitement.

Beadle, of course, succeeded Morgan as Chairman of the Caltech Biology Division. There is an interesting parallel in the careers of these remarkable figures in twentieth century biology. Morgan worked with marine animals before his monumental work with Drosophila, but when he retired he returned to working on marine animals. George Beadle's first love was maize. His work with Drosophila and then with Neurospora gave us biochemical genetics. But when he retired, Beadle went back to his first love, his interest in maize and its origin. Today he tells us how that most recent research is progressing.

Matt Meselson, our next speaker, has clearly had his share of experiencing and taking part in many exciting scientific discoveries. It may surprise some of you to know that his Ph.D. was obtained in the Chemistry rather than the Biology Division and that he was an Assistant Professor of Chemistry before becoming a Senior Research Fellow in Biology. He left Caltech to go to Harvard in 1960, where he is Cabot Professor of Natural Science. His talk today will deal with his recent work on transposable genetic elements in Drosophila. These mysterious elements may somehow be involved in spontaneous mutation and therefore could have great evolutionary significance. In any case, their high degree of variability within strains, even of the same species of Drosophila, is intriguing.

Few will know that while he was still in high school Matt had a job assisting in the care of the Drosophila stock collection at Caltech. Thus, he is now working on the organism which I like to think might have first kindled his interest in biology. It is a pleasure to have Matt back with us and we in the Division will have the added benefit of having him here for a more extended period commencing in January 1979 as a Fairchild Distinguished Scholar.

Our next speaker is Dr. S. C. Shen from the Institute of Plant Physiology, Academia Sinica, Shanghai. Dr. Shen was a graduate student under Professor Horowitz starting in 1947, and he received the Ph.D. degree from Caltech in 1951. Shen was one of the first persons to characterize tyrosinase from Neurospora. He has since worked on the genetics of antibiotic production in fungi and is currently studying the biochemical genetics of nitrogen fixation in the bacterium Klebsiella. There has been much discussion recently of the feasibility of cloning the nitrogen fixation genes and then, using recombinant DNA techniques, of transferring those genes to the many agriculturally important plants that cannot fix nitrogen.

I would like to add that Dr. C. C. Tan, who is Vice President of Fudan University in Shanghai, as well as Director of the Genetics Institute of that University, is also attending our anniversary celebration and giving lectures here on campus. Dr. Tan was a graduate student at Caltech from 1934 to 1937 working with Dobzhansky and Sturtevant on genetical and evolutionary problems in the genus Drosophila. So, it is a great pleasure to welcome back both Dr. Shen and Dr. Tan after these many years.

THE ORIGIN OF MAIZE

G. W. Beadle, Nobel Laureate

President Emeritus, University of Chicago

Corn is clearly of Western Hemisphere origin and was the most important food plant of the peoples of the pre-Columbian Americas. It played decisive roles in the rise of the great Toltec-Aztec, Maya, and Chimu-Inca pre-Columbian culture centers. As one example, at the height of the Toltec-Aztec nation, its rulers exacted tribute of the 26 conquered nearby nations. This tribute consisted of 300,000 bushels of corn per year, an equal quantity of dry beans, as well as other foods, gold, and many valuables, all carried on the backs of men, women and children for many, many miles.

Corn stolen from the Indians kept the Pilgrims from starving during their first terrible winter in New England.

Corn is today the world's third most important grain crop after wheat and rice. The genetics of corn is better known than that of any other flowering plant, in large part because of the efforts of the late R. A. Emerson of Cornell, a native Nebraskan, and also a former student and faculty member of that University. So far as I know, corn and E. coli are the only plants to harbor controlling elements of the kind found in corn by McClintock and later found by Jacob and Monod in the colon bacillus. The origin of corn has been more confused by taxonomists than that of any other major cultivated human food plant, largely because corn has no wild relative from which classical taxonomists believe it could have been derived. Thus, no other plant has been assigned by them to the genus of corn, Zea. In striking contrast, all other major cultivated human food plants have half a dozen or more plants assigned to their genera.

In 1790 a Spanish botanist, Hernandez, described a Mexican plant that he said "looks like corn but its seeds are triangular." Later taxonomists assigned this plant to a new genus called Euchlaena mexicana, because it grew in Mexico. Its most widely used common name is teosinte. Unlike corn, the female spike of teosinte has no cob, and its kernels are enclosed in triangular, heavy lignified segments that are called fruit cases. In each rank of teosinte segments, there is one kernel, while in corn the kernels are paired and are naked. Teosinte spikes are two-ranked, thus it has a two-ranked ear, whereas the simplest normal corn is four-ranked and with two kernels per rank, thus producing an eight-rowed ear. Plant anatomists and morphologists have thus been extremely skeptical about the possibility of deriving corn from such a different plant.

In the late 1880's teosinte was brought to the United States and Europe where it was tested for a forage crop. It grew luxuriously but it did not produce seed. With the discovery of photoperiodism in 1920 by Garner and Allard, Emerson set out to study the relation of teosinte to corn, and he soon found it would grow and reproduce if the daily summer photoperiod at Cornell was artificially shortened to 13 hours or less. By 1926 he had made many hybrids and as his part-time assistant I was assigned to study the hybrids genetically and cytologically. We knew both species had ten pairs of chromosomes. Cytological studies showed that the teosinte and corn chromosomes paired essentially as regularly as those in pure corn. We could also measure genetic crossing-over in the F-1 hybrids, because we had nine chromosomes with at least two genes contrasting in alleles in each. We found crossing-over was essentially the same as that in pure corn.

That seemed pretty much to settle the matter. Teosinte was an entirely reasonable ancestor of corn. That was in the early thirties and we both went off to other investigations. But alas, it didn't stay settled, for in 1938 Mangelsdorf and Reeves, then at Texas A & M, proposed an entirely new hypothesis involving a third species, <u>Tripsacum</u>, an obvious relative of maize but also clearly more distant than teosinte. Because teosinte appeared to them in many ways intermediate between corn and <u>Tripsacum</u>, they postulated that teosinte was not ancestral to anything but instead was the offspring of a hybrid between <u>Tripsacum</u> and a postulated primitive wild corn, much smaller than modern corn.

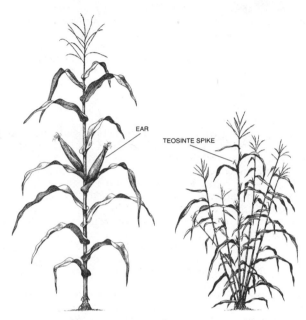

FIGURE 1 — *Comparison of maize and teosinte plants and their female spikes and individual fruits. In plant size both species vary widely, depending on genetic and environmental factors. Kernel sizes likewise vary, especially in maize. (Beadle, Scientific American 242, Jan. Copyright © 1980 by Scientific American, Inc. All rights reserved.)*

They produced a large and impressive monograph on this new hypothesis called the "tripartite hypothesis" because of the three species. I thought it entirely untenable, and said why in a 1939 paper. The <u>Tripsacum</u> they used has 18 pairs of chromosomes; corn has 10. The hybrid is difficult to make artificially and has never been observed in nature despite astronomical numbers of opportunities for it to occur. None of the 18 chromosomes of <u>Tripsacum</u> pairs normally with any one of the 10 of corn. The hybrid is sterile except for rare and reduced egg cells fertilized by corn sperms. From many such backcross-combinations and others, nothing significantly resembling teosinte has ever been observed. My opinion had zero influence--in part, because I had gone off onto other types of investigation.

Later, in 1967, shortly before my retirement from academic administration at Caltech and the University of Chicago, I received a copy of H. Garrison Wilkes's excellent monograph, "Teosinte the Closest Relative of Maize," which I read with a great deal of admiration except for his characterization of the teosinte origin of corn as a "myth," and, in another context, a "crude hypothesis." He was a student of Mangelsdorf. I then and there resolved to do something about 34 years of confusion that I attributed to the tripartite hypothesis. I now have been doing this for the past nine years, during which Wilkes and I have become good friends and have collaborated in collecting trips in Mexico and Guatemala. I hope he has abandoned his former heretical views, though he has not to my knowledge done so.

FIGURE 2 — *Six ears of tunicate teosinte from different plants showing variability.*

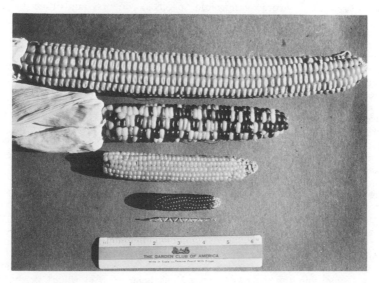

FIGURE 3 — *Four ears of corn showing some range of size and type, and teosinte ear next to rule.*

An obvious approach toward clarifying the teosinte-corn relation was to make further studies of their hybrids. This had not been done in the U.S. Corn Belt because of the complication of differing photoperiod responses among the F-2 and other segregants. To avoid this, I asked Edwin Wellhausen of the International Maize and Wheat Improvement Center in Mexico if we might grow F-2 and backcross hybrids in their test plots near Mexico City, where the photoperiod was favorable for all segregants. He generously agreed. We proposed to plan for 50,000 F-2 and backcross plants, but initially grew about 16,000, which turned out to be sufficient for our purpose. We could not have done even this without the full cooperation of a number of people in the International Maize and Wheat Improvement Center, particularly Dr. Mario Gutierrez and Prof. Walton Galinat, another student of Mangelsdorf. We determined that parental phenotypes appeared in frequencies of about 1 in 500. Because all segregants were open-pollinated, we had no way of determining how many of these were true breeding. But it seemed clear that there could not be many different major independently segregating genes in corn and teosinte.

We then decided we could refine this investigation in two different ways. Instead of using modern corn, it would be much more meaningful if we could use the most primitive known types of corn. But the most primitive types of corn are not possible because they are archaeological, 7000 years old, and obviously will not germinate. It seemed reasonable, however, that such primitive corn types could be found among segregants of hybrids between teosinte and small modern types of corn. This has proved to be so.

Finally in 1974, thirty-six years after it was proposed, Mangelsdorf conceded that the long-defended tripartite hypothesis was untenable, and in its stead he proposed that teosinte evolved from a postulated wild corn much like the earliest Tehuacan corn of some 7000 years ago. I and some others regard this new view as untenable as its predecessor because the most primitive known corns, and all intermediates between them and teosinte, have no effective seed-dispersal mechanisms. Since it could not survive in nature, the postulated wild corn could not have existed to give rise to teosinte. It is also clear that all known examples of corn, including the earliest Tehuacan specimens and all reconstructions of it, are so vulnerable to birds, rodents, and other predators that none could have survived without human intervention. We have made numerous experiments on this. As one example we have given tree squirrels in Jackson Park a choice of teosinte fruits and corn kernels. A squirrel will pick up a corn kernel and immediately eat out the germ. If he is not hungry, he will bury the emasculated kernel. If he tries a teosinte seed, he first tries to find the germ but cannot because it is on the back side of the heavy fruit case. So he buries the entire fruit, thus saving the teosinte to grow later.

In contrast to corn, teosintes with their highly effective lignin-hemicellulose fruit cases and effective means of scattering their seeds are in all respects remarkably successful wild plants.

Consistent with the full fertility of Mexican teosinte hybrids with corn, their seed storage proteins and their 19 isozymes at five genetic loci are essentially identical in the two species, which again shows their very close relation. In all these respects, the Tripsacums are very different. Dark-staining heterochromatic chromosome knob positions are likewise remarkably similar in corn and Mexican teosintes, but quite different from those of Guatemalan teosintes. This and the greater sterility of Guatemalan teosinte hybridized with corn (as compared to Mexican teosintes hybridized with corn) make it likely that corn came from Mexican teosintes rather than from those of Guatemala.

Contrary to statements in the literature, we have demonstrated that the yields of teosinte in the wild are not unlike those of wild wheat in the Near East. It has also been said that one cannot harvest teosinte seeds because they so effectively scatter their "seeds." The fact is that with nothing more than a plastic sheet, comparable to an animal skin, it is easy for one person to collect teosinte at the rate of one to two kilograms per hour. Thus a family group might well have harvested a metric ton in one season.

It has been repeatedly said that teosinte is a highly improbable human food source because its small kernels are enclosed in heavy lignified fruit cases weighing about the same as naked kernels. The fact is teosinte is used in several ways as human food, and surely was in primitive cultures. Slightly underripe whole fruits (kernels plus cases) can be eaten with no difficulty. Whole ripe seeds can be popped out of their cases by heating, and the kernels are then indistinguishable from popped corn except in kernel size. Mature dry fruits ground and pounded with primitive grinding stones are directly edible, or their kernel fragments can be separated from broken cases by water flotation requiring no equipment more elaborate than an animal skin.

Professor Mangelsdorf and I disagreed about whole ground teosinte as a plausible human food, so we appealed to Professor Scrimshaw, a nutritional authority, to serve as referee. He proposed the following experiment: I was to eat 75 grams of whole teosinte meal for each of two days. If there were no adverse symptoms, I was to eat 150 grams for each of two more days, and if there were no untoward symptoms, I had made a good case. There were none.

There is some evidence that teosinte was eaten in pre-Columbian times. Ford and Drennan of the University of Michigan have found carbonized teosinte fruits in refuse heaps near ancient dwellings near Oaxaca, Mexico. It seems reasonable to assume that it was collected for human food.

Linguistics suggest something about corn's origin. The name teosinte is derived from the Nahuatl language, teotl meaning God and centli meaning ear of corn, thus "God's ear of corn." Throughout the 1500-mile range of teosinte in Mexico and Guatemala--spanning many native Indian languages--all references to teosinte seem to have been translated to Spanish as madre de mais, mother of corn. Is this a kind of cultural memory? I believe so.

It is sometimes said by natives in parts of Mexico where small family plantings of corn are grown that "teosinte is good for the corn." In fact it is, but only through hybridization with teosinte followed by successive backcrosses to corn. This counteracts the deleterious effects of repeated inbreeding.

This rejuvenation of corn after too long inbreeding is also the basis of the not infrequent assertion by natives that "teosinte turns into corn in three years."

Modern corn pollen is significantly larger than that of teosinte. Otherwise, they are indistinguishable microscopically, but differ from all other known pollens.

Some years ago 14 pollen grains were found in a drill-core sample taken 200 feet under Mexico City and estimated to have been deposited 25,000 to 80,000 years ago. On the basis of size, five of these are judged by Barghoorn and associates to be those of corn, not teosinte. Thus it was concluded that there had existed a wild corn before man in the Western Hemisphere, and that it was ancestral to teosinte.

Since the earliest known archaeological corn types, and all others, have no means of seed dispersal and are highly vulnerable to destruction by rodents, birds, insects, and other predators, it would seem most improbable that such a postulated wild corn could have existed.

In both corn and teosinte there are positive functional correlations of pollen size and silk length through which pollen tubes must grow to effect fertilization. Large pollen on short silks is less than maximal effective use of pollen resources, while small pollen on long silks has insufficient resources to carry male nuclei through longest silks. Thus modern corn with large ears and long silks has large pollen, and small primitive types with short silks have smaller pollen.

The most primitive known corn is found in a dry cave in the Tehuacan

Valley of Mexico. It is carbon dated some 7000 years before the present and can be assumed to have been cultivated because it has no known way of scattering its kernels. Its silks were surely short, and thus its pollen must have been smaller than that of modern corn.

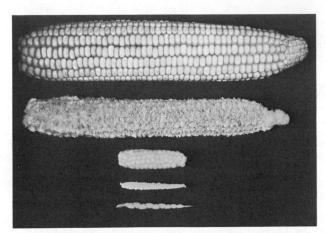

FIGURE 4 — *Top, corn belt corn and cob. Center, primitive corn reconstructed from corn teosinte hybrid, ear and cob. Bottom, teosinte ear.*

The pollen found 200 feet below the present level of Mexico City includes five grains approximately the size of modern corn pollen. On the basis of the known correlation of silk length and pollen size, this would indicate that these five grains are too large for a primitive corn of 25,000 to 80,000 years before the present. Thus other interpretations must be considered.

It is not inconceivable the large pollen could be that of a tetraploid teosinte not yet reduced by natural selection to minimal effective size. Present-day tetraploid teosinte pollens are approximately the size of those diploid counterparts, but these are presumed ancient and thus are consonant with short silk lengths.

Contamination of drill-core samples with modern pollen grains just visible to the naked eye may be improbable, but cannot be absolutely excluded because the samples under consideration were taken primarily for construction purposes. Even with all reasonable precautions, the intrusion of extraneous materials of microscopic size cannot be reduced to zero under such circumstances. This is especially so in an area where corn is grown close to Mexico City itself, where whirlwinds or dust devils are frequent and where volcanic and seismic disturbances are not unknown.

Just when and where the American Indians transformed teosinte into corn we do not know, but it was surely the most remarkable single plant-breeding achievement of all time. I believe it must have begun at least 10,000 years ago, and could have been initiated by the finding of a single dominant mutant type we now know in corn as tunicate or pod corn. This mutant type lengthens and strengthens the chaff-like glumes basal to individual kernels so as to enclose them in miniature husks. Transferred genetically to teosinte, it reduces and softens the fruit cases and reduces the tendency of the rachis segments and fruits to fall apart, thus permitting easy threshing out of naked kernels. Such a mutant could well have been a simple and decisive step in the early stages of the transformation of teosinte to corn since in one mutational step it could have become a far more desirable food plant.

Had such a mutant type been observed in a perennial diploid teosinte such as was recently found in Mexico, it could have been especially significant, because once established in a favorable genetic background, it could have been disseminated asexually more or less indefinitely with little change and no need to replant annually. Such a mutation could well have served as an incentive to the more rapid human-directed evolution of the annual teosintes on their way to corn.

For all cultivated plants except corn, there are many wild relatives that plant breeders can exploit for desirable genetic traits, but for corn there is only one, teosinte. As a favorite food of grazing animals, teosinte is an endangered species confined to Mexico and Guatemala where overgrazing is rampant.

Some of us have been collecting and seed-banking teosintes, but we should be doing much more, especially in preserving living populations. Corn breeders have no other comparable source of genetic diversity, and we think it highly important to preserve it for both practical and aesthetic reasons.

Selected Readings

Beadle, G. W. Teosinte and the origin of maize. In: Maize Breeding and Genetics. D. B. Walden, ed. John Wiley and Sons, Inc.: New York, pp. 113-128 (1978).

Beadle, G. W. Studies of Euchlaena and its hybrids with Zea. I. Chromosome behavior in E. mexicana and its hybrids with Zea mays. Zeitschrift für Abstammungs Vererbungslehre 62:291-304 (1932).

Emerson, R. A. and G. W. Beadle. Studies of Euchlaena and its hybrids with Zea. II. Crossing over between the chromosomes of Euchlaena and those of Zea. Z. Abstam. Vererbungsl. 62:305-315 (1932).

Galinat, W. C. The origin of corn. Annual Review of Genetics 5:447-478 (1971).

Mangelsdorf, P. C. In: Corn, Its Origin, Evolution and Improvement. Belknap Press: Cambridge, Mass., Harvard University, pp. 1-262 (1974).

Warren, L. O. Teosinte as a host of stored grain insects. Journal of Economic Entomology 47:630-632 (1954).

UNSTABLE DNA ELEMENTS IN THE CHROMOSOMES OF <u>DROSOPHILA</u>

Matthew Meselson, Pamela Dunsmuir, Miriam Schweber and Paul Bingham

Department of Biochemistry and Molecular Biology
Harvard University, Cambridge

Chromosome elements that can transfer to new chromosomal positions were discovered by McClintock (1950) in maize. Although not discernable under the microscope, the transposable elements are detected by their genetic and cytologic effects. These include (1) alteration of the expression of nearby genes, (2) chromosome breakage or rearrangement at nearby sites, and (3) promotion of the transposition of certain other elements on the same or different chromosomes.

The more recent discovery of transposable DNA sequences in bacteria, particularly insertion sequences and transposons, has stimulated interest in such elements and encourages the view, long propounded by McClintock, that they are of universal occurrence and importance in all organisms. In <u>Drosophila</u>, however, the higher organism most investigated by geneticists, evidence for transposing elements has only recently appeared (Green, 1969; Ising and Ramel, 1976).

We are conducting a general test of the prevalence of transposable elements in <u>Drosophila</u> and seeking to understand their behavior by making use of the methods of DNA cloning and <u>in situ</u> hybridization to chromosomes (Wensink, Finnegan, Donelson and Hogness, 1974). In order to detect possible transpositions, we ask if particular DNA sequences cloned from <u>Drosophila</u> occur at noncorresponding sites on the chromosomes of individuals from two closely related <u>Drosophila</u> species or of different individuals of the same species. Thus, we can detect sequence transpositions which may have occurred over a broad time scale. The cloned DNA sequences tested are 1000 to 10,000 base pairs long, a size range like that found for the transposons of <u>E. coli</u>. Hybridization <u>in situ</u> permits us to determine where the cloned sequence and sequences homologous to it occur on the chromosomes. This involves affixing the chromosomes to a microscope slide and incubating them with tritium-labeled cloned DNA under conditions that allow chains of the labeled DNA to form molecular hybrids with complementary chains of chromosomal DNA. After washing away unbound radioactive DNA, the slide is covered with a thin layer of photographic emulsion, kept for a few weeks and then developed. Under the microscope one sees clusters of silver grains superimposed on the chromosomes where the radioactive DNA has hybridized to chromosomal sequences.

Ordinary chromosomes have too little DNA and are too short to provide the necessary sensitivity and resolution. Instead, polytene chromosomes, formed in certain tissues as a result of many consecutive duplications without segregation, are used in this procedure. Figure 1 shows the polytene chromosomes from a salivary gland nucleus of a late larva of <u>D</u>. melanogaster. The pattern of cross-bands is highly reproducible, permitting precise identification of corresponding locations on similarly prepared polytene chromosomes of different individuals. In these polytene nuclei, the chromosomes are fused together at their centromeres, giving rise to a chromocenter with chromosome arms radiating from it. The maternal and paternal homologues in each arm are closely paired, except for occasional regions of asynapsis, two of which occur in the nucleus shown.

Figure 2 shows a portion of the X chromosome from a nucleus of a hybrid larva produced in a cross between <u>D</u>. melanogaster and the closely related sibling species <u>D</u>. simulans. The pattern of chromosome banding in the two species is almost the same. The hybrid, although sterile, is viable and allows comparison of the pattern of hybridization on the chromosomes of the two species to be carried out in the same nucleus, insuring identical experimental conditions. In this example, hybridization in situ was done with a 7 kilobase DNA segment cloned from D. simulans. It labels only one site, the same in both species--an observation facilitated by local asynapsis in the nucleus shown. This behavior is found for 15 of 23 unselected clones, 8 from melanogaster and 7 from simulans. That is, label is found at only one site in the species of origin and only at the same site in the sibling species. Moreover, there is no significant suggestion of asymmetry in labeling intensity between asynapsed homologues in the nuclei of hybrid larvae. The 15 sites are all located in different places, with some on each of the five long chromosome arms.

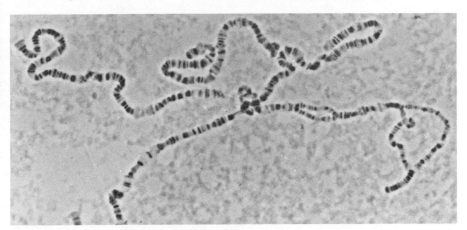

FIGURE 1 — *Polytene chromosomes from a larval salivary gland nucleus of D. melanogaster. One chromosome arm (X) is broken away from the chromocenter and one arm (2R) is asynapsed in two regions.*

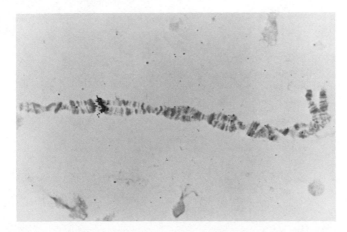

FIGURE 2 — *Hybridization in situ by a cloned DNA sequence homologous to only one chromosomal site. The 7 kilobase sequence cloned from D. simulans labels subdivision 8E on both homologues of the X chromosome of a melanogaster-simulans hybrid larva.*

The positional stability of these 15 DNA sequences is in conformity with the close similarity of the patterns of chromosome banding in the two species (Horton, 1939). It also parallels the finding that corresponding genes occupy homologous positions on the respective genetic maps of melanogaster and simulans, as determined by recombination studies (Sturtevant, 1929).

Such stability is not exhibited, however, by cloned DNA sequences that hybridize at multiple chromosome locations in the species of origin. This is illustrated in Figure 3, which shows patterns produced by a 6 kilobase segment of DNA cloned from D. melanogaster hybridized to the chromosomes of melanogaster (3a) and simulans (3b). At least 25 sites plus the chromocenter are labeled in the former, but only the chromocenter and three sites are labeled in the latter. The comparison of labeled sites is facilitated by the use of nuclei from hybrid larvae as shown in Figure 3c. When the homologues are very closely synapsed, however, it is difficult to find sites labeled on only one side, an effect we are at present studying.

Each of the eight unselected multi-site clones--five from melanogaster and three from simulans--shows very extensive species differences in its pattern of chromosome labeling. Although some sites are common, most are not; some are labeled only on the chromosomes of the species of origin and some only on the chromosomes of the sibling. At this stage our analysis is not complete enough to allow more detailed conclusions regarding the distribution of sites.

We find that the pattern of hybridization in situ by multi-site clones differs not only between melanogaster and simulans but, to a lesser though still considerable extent, even between individuals of the same species, as has been reported by Ananiev, Gvozdev, Ilyin, Tchurikov and Georgiev (1979) using clones selected for homology to RNA abundant in cultured Drosophila cells. In further agreement with these authors, we have not observed differences among nuclei from the same individual.

We think it unlikely that the extensive differences we see between and within species in the hybridization patterns of multi-site sequences result from sequence divergence, since no such differences are found for sequences that hybridize at only a single site. Also, as has been noted above, some of the multi-site clones hybridize at sites in the sibling species which are not labeled in the species of origin. Neither do we consider that localized differential replication in polytene chromosomes can account for our observations. This is because cloned multi-site sequences that hybridize at many more sites in one species than in its sibling are found to hybridize with corresponding inequality to DNA extracted from embryos of the two species and also to the nuclei of imaginal disc cells, which are not polytene.

The picture that we think fits our results best is that most of the unique DNA sequences of D. melanogaster, thought to comprise some 70% of the DNA of the genome, have remained in homologous positions since melanogaster and simulans diverged. In contrast, dispersed moderately repeated DNA sequences, presumably corresponding to the several kilobase long elements comprising approximately 15% of the melanogaster genome (Manning, Schmid and Davidson, 1975), usually do not remain in fixed positions over such intervals of time. Since we have detected only the instability of dispersed repeated elements and not their actual transposition from one site in a chromosome arm to another, we will refer to such elements as unstable rather than transposable. It should be noted that one of their apparent chracteristics, namely different abundance in related species, has been reported for certain dispersed repeated sequences in sea urchin DNA (Moore, Schiller, Davidson and Britten, 1978).

Nothing definite is known regarding the possible functions served by the unstable repeated elements. We have, however, studied one such element that is transcribed to give an abundant RNA in response to a specific stimulus, namely temperature elevation. This is a 1.5 kilobase sequence cloned from the

90

(a)

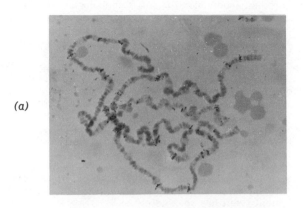

(b)

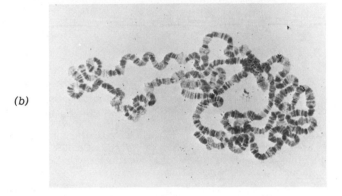

(c)

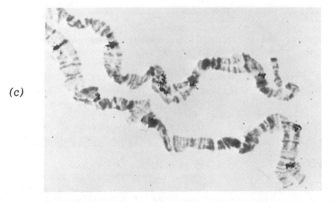

FIGURE 3 — *Hybridization in situ by a cloned 6 kilobase melanogaster sequence homologous to many sites in melanogaster and to three sites plus the chromocentral region in simulans. (a) melanogaster; (b) simulans; (c) partly asynapsed portions of the X chromosome and the right arm of chromosome 2 from a melanogaster-simulans hybrid larva. Sites labeled on only one homologue are evident.*

heat shock locus in chromosome subdivision 87C of D. melanogaster, which also hybridizes in situ at several other chromosomal sites, including the chromo-central region (Lis, Prestidge and Hogness, 1978; Livak, Freund, Schweber, Wensink and Meselson, 1978). In D. simulans, however, this element hybridizes only in the chromocentral region and no RNA homologous to it is detected after temperature elevation.   D. melanogaster homozygous deficient for a short region including the 87C heat shock locus also fails to produce RNA homologous to the 1.5 kilobase sequence.   Nevertheless, the induction and subsequent regression of puffs at the five other major heat shock loci and the production of the seven major heat shock proteins regularly seen in sodium dodecylsulfate gel electrophoretic analyses, as well as the general attenuation of other protein synthesis which results from temperature elevation all appear unaffected by the deficiency. Neither does the cloned 1.5 kilobase sequence exhibit homology to any heat shock messenger when tested by hybrid arrested translation in vitro. Thus, the 1.5 kilobase element belongs to the class of unstable elements and is abundantly transcribed in response to a specific stimulus, but no function for either the RNA or the element itself has been detected.

Robert Freund and Aaron Perlmutter conducted studies of heat shock protein synthesis in D. melanogaster deficient for 87C1. Work described in this report was supported by the National Institutes of Health.

Selected Readings

Ananiev, E. V., V. A. Gvozdev, Yu.V. Ilyin, N. A. Tchurikov and G. P. Georgiev. Reiterated genes with varying location in intercalary heterochromatin regions of Drosophila melanogaster polytene chromosomes. Chromosoma 70:1-17 (1978).

Bridges, C. B. Salivary chromosome maps with a key to the banding of the chromosomes of D. melanogaster. Journal of Heredity 26:60-74, (1935).

Green, M. M. Controlling element mediated transpositions of the white gene in Drosophila melanogaster. Genetics 61:429-441 (1969).

Horton, I. H. A comparison of the salivary gland chromosomes of Drosophila melanogaster and D. simulans. Genetics 24:234-243 (1939).

Ising, G. and C. Ramel. The behavior of a transposing element in Drosophila melanogaster. In: The Genetics and Biology of Drosophila. M. Ashburner and E. Novitski, eds. Academic Press, Vol. 1b, pp. 947-954 (1976).

Lis, J. T., L. Prestidge and D. S. Hogness. A novel arrangement of tandemly repeated genes at a major heat shock site in D. melanogaster. Cell 14:901-919 (1978).

Livak, K. F., R. Freund, M. Schweber, P. C. Wensink and M. Meselson. Sequence organization and transcription of two heat shock loci in Drosophila.   Proceedings of the National Academy of Sciences, Washington, 75:5613-5617 (1978).

Manning, J. E., C. W. Schmid and N. Davidson. Interspersion of repetitive and nonrepetitive DNA sequences in the Drosophila melanogaster genome. Cell 4:141-155 (1975).

McClintock, B. The origin and behavior of mutable loci in maize. Proceedings of the National Academy of Sciences, Washington, 36:344-355 (1950).

Moore, G. P., R. H. Scheller, E. H. Davidson and R. J. Britten. Evolutionary change in the repetition frequency of sea urchin DNA sequences.   Cell 15:649-660 (1978).

Sturtevant, A. H. The genetics of Drosophila simulans. Carnegie Inst. Wash. Publ. 399:1-62 (1929).

Wensink, P. C., D. J. Finnegan, J. F. Donelson and D. S. Hogness. A system for mapping DNA sequences in the chromosomes of Drosophila melanogaster. Cell 3:315-325 (1974).

GENETICS OF BACTERIAL NITROGEN FIXATION

S.-C. Shen

Institute of Plant Physiology, Academia Sinica
Shanghai, People's Republic of China

Biological nitrogen fixation is governed by the crucial enzyme nitrogenase, the necessary electron-transfer and energy-generating system that catalyzes the synthesis of NH_4^+ from atmospheric nitrogen gas. All the elements for nitrogen fixation are specified by a set of nitrogen-fixation genes. Investigations on the nature and manipulation of nitrogen-fixation genes of bacteria has been proceeding with great rapidity in recent years. I would like to give a brief account of our work on the nif genes and their regulation in Klebsiella pneumoniae.

Fine structure mapping of nif genes of Klebsiella pneumoniae

A set of nif mutations linked with the histidine operon has been demonstrated in several laboratories, using transducing phage P1 or by conjugation mediated by an R factor derived from E. coli. Further insight into the location of the nif genes relative to other genetic markers on the bacterial chromosome was demonstrated using deletions in the histidine region.

Complementation analyses using a P group plasmid pRD1 among various nif mutations were performed for delineating the nif genes. Dixon and Kennedy (1977) first mapped out the order of nine nif genes aligned on the chromosome of K. pneumoniae. They pointed out that these nif genes are divided into two clusters: a his-proximal cluster and a his-distal cluster. These two clusters are separated by a region of 9 kb in length; Kennedy named this region a "silent" region. To reexamine their results, we reanalyzed the order of nif mutations. A more reliable method for studying the percentage of co-transduction was adopted. A given hisD mutant, C-H80, was used as the common recipient and various nif mutants to be mapped were used as the donors in P1-transduction. We found that the maximal difference of co-transduction percentage is only 6%.

The co-transduction data for nif genes are analyzed according to the equation $f = (1-d/L)^3$, where f = co-transduction frequency and L = length of transducing DNA, here taken to be 80 kb. Calculations of "d" for the distances between hisD and nif mutation sites were based on the mean transduction frequencies for each mutation. Our results indicate that about 13 kb length of DNA separates the mutation N-213 from hisD, the most his-proximal mutation site. These data are comparable with those of Dixon et al. But we didn't find any big gap as shown by Kennedy among these nif mutants. The physical distance between the two most distant mutations is about 1-2 kb. Our results have been confirmed recently by both the Wisconsin and Sussex groups.

Complementation analysis of nif mutations

The method used for complementation test was mainly developed by Dixon (1977). The nif⁻/nif⁻ heterogenotes were constructed by transferring nif plasmid mutant pRD-nif⁻ from E. coli JC5466 to various K. pneumoniae nif⁻ recipients. Complementations were obtained, but the complementation patterns were complex. This may be due to presence of multiple mutations in our nif mutants, which were obtained by mutagenesis with nitrosoguanidine, and the occurrence of recombination in a perodiploid. However, we will have identified about 7 cistrons among these nif mutants.

Brill et al. (1978) have isolated several hundred nif⁻ mutants that were obtained by different mutagens, e.g., nitrosoguanidine, phage mu, diethylsulfate, hydroxylamine, etc. Before complementation tests, they carefully made the recipient strain rec⁻ so that complementation could be distinguished from recombination in mating experiments; likewise in these experiments merodiploids containing a nif mutation in the chromosome and nif mutation in the plasmid were constructed. Fourteen nif genes were identified. The polarity and number of operons were determined by complementation patterns of strains with a mu insertion.

Complementation experiments were performed on polar mutations in each nif complementation group, resulting in the assignment of 14 genes to 7 operons. The functions of these genes were determined by preparing active extracts from various nif mutants and then making different combinations of these extracts, or by adding pure component I, component II or the iron-molybdenum cofactor to the extracts to reconstitute acetylene reduction activity.

Up to the present we can be certain that nifK, nifD, and nifH determine the structure of three nitrogenase polypeptides and that nifB, nifN, nifE, and nifF determine the iron-molybdenum cofactor and electron transport factor, respectively. It was speculated that nifA and nifL are required for the expression of all of the nif genes and that nifK, nifD, and nifH, besides specifying the component I, component II polypeptides, help to regulate the synthesis of all of the nif proteins.

Recently we found that nif mutant N-120 which was mapped on the distal side of histidine operon shows a different function in chemical complementation tests. Extracts prepared from mutant N-120 fail to reconstitute the acetylene reduction activity after mixing them either with the purified component I of nitrogenase or with the extracts of some component I defective mutants as the source of component II. According to immunoelectrophoretic tests of the nif proteins of mutant N-120, we found that component I and component II of nitrogenase are markedly reduced in amount in comparison with the wild type. The N-120 mutation apparently results in depressing the synthesis of all the components of nitrogenase. Since the addition of purified component I to the extract of nifJ mutant 1660 can bring about the reconstitution of nitrogenase activity, it indicates that the cistron J mutant behaves as a component I mutant. The mutant N-120 appears to be different in function from the cistron J and presumably represents another nif gene for nitrogen fixation.

Regulation of nif genes

Glutamine synthetase is thought to play an important role in regulating nitrogenase synthesis and has been indicated in a number of experiments. We have isolated many hisD-unlinked nif⁻ mutants; some of them apparently do not belong to the category of glutamine synthetase mutants, because they are not glutamine auxotrophs and are all provided with glutamine synthetase activity. One of such nif⁻ mutants, C-7, possesses the glutamine synthetase activity of about 60% of the wild type, but its glutamate synthetase activity corresponds to only about 30% of the wild type. It is able to grow on excess NH_4^+; presumably glutamate dehydrogenase, instead of glutamate synthetase, is used here for the synthesis of glutamate.

When mutant C-7 is spontaneously reverted to the wild type, or the transductants his⁻ nif⁺ are obtained by crossing it with the wild type, their glutamate synthetase and glutamine synthetase activity completely resume the normal level accompanying the recovery of the activity of nitrogenase. The results of immunoelectrophoretic analysis showed that both mutants C-5 and C-7 contain only trace amounts of components I and II of nitrogenase in extracts. When extracts of these mutants are mixed with that of the mutant N-120, since they are all lacking nitrogenase components, no recovery of the nitrogenase activity can be demonstrated. Further studies on the mechanism of their function have been made in our laboratory.

As mentioned before, many nif genes, besides specifying their own products, exert also a regulatory effect on other nif genes. In view of the complexity of the interaction of nif genes and the complexity of their metabolic control, difficulties encountered in cloning the nif genes for suitable expression in a host can be anticipated.

Selected Readings

Brill, W. J. Nitrogen fixation: basic to applied. American Scientist **67**:458-466 (1979).

Hsueh, C.-T., J.-C. Chin, Y.-Y. Yu, H.-C. Chen, W.-C. Li, M.-C. Shen, C.-Y. Chiang and S.-C. Shen. Genetic analysis of the nitrogen fixation system in Klebsiella pneumoniae. Scientia Sinica **20**:807-817 (1977).

Streicher, S. L. and R. C. Valentime. Comparative biochemistry of nitrogen fixation. Annual Review of Biochemistry **42**:279-302 (1973).

SESSION IV: BIOLOGY OF CELLS

Introduction

Leroy E. Hood

Ethel Wilson Bowles and Robert Bowles Professor of Biology, Caltech

This session, the biology of cells, spans a fascinating range of topics including animal hormones, the vertebrate immune response and its regulation by major histocompatibility complex, and morphogenesis.

Our first speaker, Gordon Sato, started his scientific career as a phage geneticist with Max Delbrück. After struggling with the activation of T4 phage by urea and guanidine and other related topics, Gordon moved, successively, to Brandeis University and U.C. San Diego, where he is now Professor of Biology. At the same time, Gordon's interests shifted from the activation of bacteriophage by various kinds of chemicals to the activation of animal cells by various types of hormones. He is acknowledged as one of the world's leaders in the field of hormones and tissue culture. It is a pleasure to welcome Gordon here to discuss "Hormones and tissue culture: basic and health-science aspects."

Donald Shreffler, our second speaker, pursued a fundamental analysis of regulatory factors in the immune response throughout his professional career. Don was a Ph.D. student with Ray Owen from 1958 to 1962 and, accordingly, overlapped my own stint as a Caltech undergraduate. After postdoctoral training, Don moved to the University of Michigan and later to Washington University School of Medicine, where he is now Chairman of the Department of Genetics.

Don's career, in a scientific sense, began with his teaching of a project laboratory in Ray Owen's immunology course. This course, still extant, joins together individual graduate teaching assistants with three or four undergraduates to study in depth one particular research project. Don and one group of students discovered an interesting serum protein polymorphism in mice, denoted Ss (serum substance). Don went on to demonstrate that Ss is linked to the major histocompatibility complex (MHC) of mice--a genetic region on the 17th chromosome that plays an integral role in regulating immune responses. In subsequent years, Don used the Ss genetic marker as an invaluable aid in analyzing the genetic fine structure of the MHC. More recently, Don has demonstrated that Ss is actually one of the complement components--and thus encodes one of the effector areas of the immune response. Don will speak to us on a "Genetic and functional dissection of the major histocompatibility complex."

Our final speaker, Dale Kaiser, was a member of the Delbrück phage group during his graduate career at Caltech. After Caltech, Dale moved to microbiology at Washington University School of Medicine, and in 1966 he was a part of the massive Kornberg migration that created the Biochemistry Department at Stanford Medical School--where Dale continues as a professor. Dale's career is sharply divided into two halves. Early on, his interests focused on the genetics of lambda bacteriophages--an area in which he has made very substantial contributions. More recently, the major focus of his research has shifted from phage genetics and nucleic acid chemistry to the biochemistry of morphogenesis. Dale will tell us today about "Simple social cells."

97

HORMONES AND TISSUE CULTURE:
BASIC AND HEALTH-SCIENCE ASPECTS

Gordon H. Sato

Department of Biology
University of California, San Diego

I would like to tell you why I feel so keenly the great privilege it is to be addressing this symposium:

I grew up in a tough, polyglot neighborhood in the heart of the harbor of Los Angeles. Neither my classmates nor I had any idea of academic achievement, and so in 1950 I found myself pursuing the traditional occupation of immigrant Japanese--that is, I was mowing lawns in the vicinity of Caltech. I would often drive by in my truck and wistfully think how wonderful it would be to be a student there. One day after a rainstorm, I slipped off my truck and sprained my ankle. Since I couldn't work I suddenly had enough free time to come to Caltech to try somehow to gain admission. The first man I saw was the Dean of Admissions. He quickly determined that I had been a terrible student. He said, "We only take the very best students here, but tell me what you are interested in." I said I was interested in active transport across biological membranes. He said, "That's biology--go see a man named Beadle." So I went to the Biology Department and told Dr. Beadle the same story. He told me that what I was interested in was biophysics and that I should go to see Max Delbrück.

I found Max sitting in semi-darkness in his old yellow chair--apparently deep in thought, with his office door ajar. When I knocked, he said with some annoyance, "What do you want?" When I told him I wanted to be a student, he said, "Tell me the story of your life." The long and short of it was that I was admitted as a special student with probationary status, and eventually I managed to get a Ph.D. Finding Max that day was one of the most crucial and fortunate events of my life.

In coming to Caltech, I saw honest scholarship for the first time ever. I met fellow students like Gerry Terres, Bob Metzenberg, Bruce Ames, and Dale Kaiser. I had never known such a collection of people before. As you can imagine, Caltech was an eye-opening and inspiring experience for me, for which I will be ever grateful.

The work of my laboratory has had a very simple theme: to use tissue culture as a tool to analyze and understand integrated physiology. The problem with this simple idea was that at the time this work was started there were no specialized cell cultures in existence. There were no muscle cells that contracted, no nerve cells that developed action potentials, and no glandular cells that secreted hormones in culture. In fact, all cultured cells looked like fibroblasts. Our first efforts were therefore directed toward developing differentiated cell cultures. At that time, it was believed that the deficiency of differentiated characteristics in cultured cells was due to dedifferentiation. According to this notion, some mysterious aspects of culture conditions caused cells growing in culture to gradually lose their differentiated properties (Figure 1).

I was fortunate in having a Caltech microbiological background so that I could easily entertain an alternative explanation, namely the selective overgrowth of fibroblasts. According to this hypothesis, if one cultures cells taken from a tissue such as the liver, which consists primarily of parenchyma (the specialized cells of the tissue) but includes some connective tissue fibroblasts (which are also present in almost all tissues), the fibroblasts outgrow the parenchyma so that the final culture population is fibroblastic.

98

We performed experiments which showed that the main reason for the lack of differentiated properties in cultured cells was indeed due to the overgrowth of the ubiquitous fibroblast (Figure 1). Today this is considered a truism. Everybody knows that attempts to establish cultures of the distinctive epithelial cells of a tissue are often thwarted by the selective overgrowth of the fibroblasts of the tissue of origin. However, at the time our experimental results were reported, such an idea was considered heresy, and I was subjected to severe criticism for propounding such nonsense. The opposition to the selective overgrowth hypothesis was so vehement that I received a letter from the American Cancer Society saying that I shouldn't bother to apply for a renewal of a grant.

DEDIFFERENTIATION

SELECTION

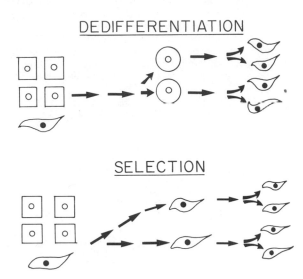

FIGURE 1 — This is a graphical representation of the two hypotheses to explain why all cell culture stains previous to the 1960's had the characteristics of fibroblasts. In this case, liver is taken as the illustrative example. The groups of cells on the left represent the cell population taken from the liver to be placed in culture. It consists mainly of the parenchyma or organ-specific cells of the liver (square cells) and to a small extent of connective tissue fibroblasts (elongated cells).

Both hypotheses explain why the final tissue culture population (group of cells on the right) are fibroblastic. According to the dedifferentiation hypothesis, parenchymal cells growing in culture are gradually converted into fibroblasts. According to the selection hypothesis, the minor component of fibroblasts in the initial inoculum outgrow the parenchymal cells so that the final culture population consists of fibroblasts derived from pre-existing fibroblasts. To determine which explanation was correct, antisera against parenchyma and antisera against fibroblasts were prepared. Cells, freshly isolated from the liver, were treated with the antisera for brief periods of time, and then placed in culture. The dedifferentiation hypothesis predicts that killing the parenchyma of the inoculum with antisera would block subsequent culture growth because it is the parenchyma which grow and become fibroblasts. The selection hypothesis predicts that killing the fibroblasts, which are only a minor component of the initial population, would block subsequent growth because it is these cells that generate the final population. The experimental results were consistent with the selection hypothesis.

It was now clear to me that the solution to the problem of obtaining differentiated cell cultures was simply to grow the desired cells out of a mixture of cell types. Again, the Caltech experience proved to be useful. As a teaching assistant for Renato Dulbecco, I was exposed over and over again to the concept of enrichment culture in microbiology. A microbiologist used this technique to grow specific micro-organisms from a mixture. For example, if a sample of soil, which contains many types of bacteria, is placed in an airtight jar of broth, only anaerobes will grow. If vinegar is added to the growth media, only acid-forming bacteria will survive and grow.

Dr. Vincenzo Buonassisi and I set out to devise an analogous enrichment culture method for animal cells. We turned to Dr. Jacob Furth, one of the important pioneers in the field of endocrine oncology. He kindly provided us with hormone-secreting tumors that were serially transplantable in mice. We placed these tumors in culture for short periods of time and injected animals with the cultured cells to obtain culture-derived tumors. The culture-derived tumors were much easier to grow in culture than the original tumors and pure strains of hormone-secreting and hormone-responsive culture lines could be established from them readily (Figure 2). In this way we established many

ENRICHMENT CULTURE TECHNIQUE

for

CULTURING DIFFERENTIATED CELLS

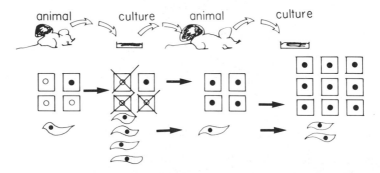

FIGURE 2 — *The method of alternately passaging differentiated tumor cells through animals and culture is graphically presented. The original tumor cell population is represented by the group of cells on the left. It consists of the differentiated tumor cells (square cells) and fibroblasts. When this cell population is put into culture, the fibroblasts proliferate to a great extent while the tumor cells die. A few exceptionally hardy tumor cells (represented by cells with dark nuclei) survive the initial culture period. When the cultures are injected into animals, a tumor is generated from these hardy tumor cells which have been selected for their ability to survive the culture conditions. The fibroblasts contribute little to the growth of the tumor because they are of non-cancerous cells. The culture-derived tumors grow more readily in culture than the original tumor because of the previous selection. From these cultures, it is now easier to isolate pure strains of functionally differentiated tumor cells. In this way adrenal cortical (steroid-secreting, ACTH-responsive), pituitary (ACTH-secreting), pituitary (prolactin- and growth hormone-secreting), Leydig (steroid-secreting), glial (S100 protein-containing), neuroblastoma (electrically-active, neurotransmitter-synthesizing), teratoma (capable of differentiation), and melanoma (pigmented) cell lines were established.*

100

functional cell lines such as steroid-producing adrenal cells, ACTH-secreting pituitary cells, growth hormone–secreting pituitary cells, pigmented melanoma cells, steroid-producing Leydig cells, teratoma cells capable of differentiation in vitro, glia and neuroblastoma cells with distinctive properties of nervous tissue, etc. We used the somewhat involved technique to give us the best chance for success because it was not known at the time whether or not it was possible to grow functionally differentiated cells. Today it is commonplace to culture functionally differentiated cells, especially from tumors, without the elaborate procedure of alternate animal and culture passage.

If tissue cultures are to be useful tools for studying animal physiology, I next reasoned that they should exhibit growth responses to trophic hormones as do the parental tissues in vivo. For instance, thyroid cells proliferate in response to thyroid–stimulating hormone, and uterine endometrial cells proliferate during the menstrual cycle under the influence of estrogenic hormones. Ten years ago, no such cell culture strains were in existence. Accordingly, we set out to develop ovarian cell cultures whose growth would be dependent on the presence of gonadotrophins. Dr. Jeffrey Clark initiated this work in my laboratory and soon obtained the puzzling result that adding gonadotrophins to the culture medium had no effect on the growth of the cells. This was unexpected because we knew that if these cells were injected into animals, they would only grow if the transplant was artificially provided with high levels of hormones.

We realized that the serum component of the medium was providing hormones and decided to use serum selectively depleted of hormones. Drs. Katsuzo Nishikawa, Hugo Armelin, and David Sirbasku devised procedures for depleting serum of hormones and these have now become standard practice. They extracted serum with charcoal to remove thyroid and steroid hormones, and passed the serum through a column of carboxy methyl cellulose to remove basic peptides with growth-promoting properties. When depleted serum was used, the ovarian cells would not grow unless hormones were added to the medium. Optimal growth required the addition of multiple hormones.

Another surprise came when we discovered that the crude luteinizing hormone (one of the gonadotrophins) we were using was providing novel basic peptides which were the active components of the hormone preparations and that pure luteinizing hormone was inactive. This led to our discovery of ovarian growth factor and fibroblast growth factor, which were eventually purified in the laboratory of Denis Gospodarowicz.

The pattern of results coming from these studies, and a brief but memorable conversation with Gordon Tomkins, caused me to wonder if the role of serum in cell culture media might simply be to provide cells with complexes of hormones. Serum or its equivalent has been an obligatory component of media since tissue cultures were first initiated by Ross Harrison over seventy years ago. No adequate explanation had ever been given for this requirement. If this surmise about the role of serum was correct, it should be possible to replace serum with complexes of hormones. This was a very exciting prospect because its implications would have far-reaching consequences for many branches of biology.

Dr. Izumi Hayashi, who was a graduate student in the laboratory, quickly obtained evidence in support of the hypothesis. She found that the serum in media used to culture GH_3 cells (a growth hormone–secreting pituitary cell line that we had established from a Jacob Furth tumor) could be replaced with a mixture of insulin (pancreas), transferrin (liver), triiodothyronine (thyroid), parathyroid hormone (parathyroid), TSH-releasing hormone (hypothalamus), fibroblast growth factor (pituitary), and somatomedin C (liver) (Table 1). I have indicated the organ sources of these substances in parentheses. All of these substances are not hormones. In fact, our hypothesis has now been extended to state that the serum in tissue culture medium can be replaced by substances from four classes:

(1) Classical hormones such as insulin and hydrocortisone.

(2) Growth factors which we consider to be hormones but whose physiological role has not yet been determined. Substances of this class readily lend themselves to discovery by cell culture methods and the name "growth factor" is more a reflection of the assay system for their detection than of their actual role in physiology.

(3) Transport proteins. So far only transferrin has proven useful but it is required by almost all cells.

(4) Attachment factors, such as fibronectin and collagen. These substances are often necessary for the proper attachment of cells to the culture vessels and probably reflects the usual relationship of epithelial cells to a basement membrane in vivo.

HORMONAL REQUIREMENTS
of

GH_3 { Growth Hormone
Prolactin Secreting
Pituitary Cells

ORGAN SOURCE

1. INSULIN	Pancreas
2. TRANSFERRIN	Liver
3. THYROXIN	Thyroid
4. PARATHYROID HORMONE	Parathyroid
5. THYROID STIMULATING HORMONE RELEASING HORMONE	Hypothalamus (Brain)
6. FIBROBLAST GROWTH FACTOR	Pituitary
7. SOMATOMEDIN C	Liver

TABLE 1 — *The hormonal and factor requirement of GH$_3$ pituitary cells for growth in serum-free medium are presented here with the organ source of each component in parentheses. In view of the immense complexity of serum, it is surprising how so few a number of defined substances can replace serum. It is also surprising that the cells give a positive growth response to so many hormones. The serum-free technology is also a powerful tool for revealing hormonal dependencies. The organs listed in parentheses are the organs that would have to be ablated to show these responses by classical endocrine experiments.*

The data of Dr. Hayashi are remarkable for several reasons. The first is that serum can be replaced by such a small number of substances. In view of the immense complexity of serum, one would not have predicted that serum could be replaced by seven defined substances for GH_3 cells. It is remarkable, on the other hand, that the cells give a positive growth response to so many hormones. This surely means that the hormonal regulation of each cell type in vivo is much more complex than is indicated by classical endocrine studies. In fact, the most challenging aspect of these findings is how the metabolic control by so many hormones of each cell in the body can be fitted into the pattern of integrated physiology. Finally, the data demonstrate the immense power of the serum-free technology for uncovering hitherto unsuspected endocrine relationships. Removing serum from cells in culture is the most radical endocrine ablation imaginable. To demonstrate the hormonal dependencies of GH_3 cells revealed by the serum-free technology by classical means, one would have to perform pancreatectomy, hepatectomy, hypophysectomy, thyroidectomy, parathyroidectomy, and brainectomy. Clearly, such experiments are impractical and one can predict with fair certainty that in the future the discovery of physiological hormone responses will be made with the use of these culture techniques.

The results obtained with GH_3 cells have now been extended to a sufficiently large number of cells to enable us to say that all animal cells with the potential for growth can grow in a serum-free, defined medium containing substances from the four classes mentioned above. Each cell type has a different and unique set of hormonal requirements, and in many cases growth in the defined media is superior to growth in serum-based media. If a key role of serum in culture is to provide complexes of hormones, it is understandable why the defined, hormone-supplemented media can be superior to serum-based media. Serum is toxic to cells in culture and so is never used undiluted. It is usually diluted to a concentration of about 10% v/v. This intrinsic toxicity is probably due to the fact that many potentially cytotoxic substances in serum do not usually leave the vascular spaces and come into direct contact with cells.

Cells whose proliferation is driven by hormones usually require them at concentrations greater than those normally found in 100% serum. In addition, there are many specialized areas of the body where certain hormone concentrations are much higher than that found in the general circulation. For these reasons, dilute serum cannot be expected to provide the ideal, specialized environment required by many kinds of cells, and this probably explains why such cells have never been established in culture using serum-based medium. Use of hormone-based media should now make it possible to establish cells in culture that have not been cultured before. One such example already exists. Drs. Saverio Ambesi and Hayden Coon at the National Institutes of Health have used hormone-based media to successfully culture normal thyroid cells with their full array of functional characteristics. Such cells had not been cultured before, despite years of strenuous effort using serum-based media. In the near future, it should be possible by these methods to establish a great variety of cell culture lines with potential clinical utility, such as insulin-secreting pancreatic cells to be used as glandular replacements for diabetics.

My main concern over the next few years will be to apply these ideas to the problem of cancer. Many of the cells we have studied are tumor cells. When tested in serum-free media, they have complex hormonal requirements. This means that the hormonal regulation of tumors may be nearly as complex as the regulation of normal organ systems. Using serum-free technology, it should be possible to develop a detailed knowledge of the endocrine physiology of tumors. I feel certain that a basic approach of this kind to cancer is the best long-term hope for useful advances in therapy.

Selected Readings

Barnes, D. and G. Sato. Serum-free growth of cells in culture. Analytical Biochemistry, submitted for publication (1979).
Sato, G., G. Augusti-Tocco and M. Posner. Hormone-secreting and hormone-responsive cell cultures. Recent Progress in Hormone Research 26:539 (1970).

GENETIC AND FUNCTIONAL DISSECTION OF THE
MAJOR HISTOCOMPATIBILITY COMPLEX

Donald C. Shreffler, Chairman

Department of Genetics
Washington University School of Medicine, St. Louis

Like the other speakers in the past few days, I just want to say I really feel very privileged to have had the opportunity to work here, and particularly privileged to have had the opportunity to work with Ray Owen. I came here as a special student because of a background in agriculture. This was an important transition, and I'm tremendously grateful to Ray for helping me to make it.

The story I'd like to tell began in the basement of Kerckhoff with some of my thesis research, which rather by chance led me into an interest in the major histocompatibility complex (MHC) of the mouse--the H-2 gene complex--and its genetic organization. In 1960 when I began this work, H-2 was a pretty obscure and exotic subject. I suspect that many of you will think it still is obscure and exotic. But since 1960 the major histocompatibility complexes have become a focal point in mammalian genetics; first, because of their role in transplantation (though I won't have much to say about that here), and more recently in immune mechanisms and in disease susceptibility which is obviously the area of current primary interest. In the near future I think we will see that this is also a very important system in molecular genetics because of the potential that the growing definition of the system offers.

What I'd like to do is give you a general overview from the perspective of our own interests--our own work--which is not necessarily always the most definitive work in the area, but which will provide a sort of semihistorical thread for the development of the understanding of the genetic organization and the varied genes and functions of the complex.

It is known that there are major histocompatibility complexes in most, perhaps all, higher vertebrates. There is a group of at least 10 MHCs defined in mammalian species; other key species are the chicken and two species of amphibians. The occurrence of MHCs in amphibians and in the avian branch of vertebrate evolution suggests that these complexes had evolutionary origins at least as early as the higher amphibians. A key aspect of these systems is the substantial degree of homology that's seen among the varied genes, products, and functions associated with them. They are also characterized by a high degree of genetic polymorphism.

As I said, these major histocompatibility complexes have a number of common features in their functional organization. Figure 1 shows the four best defined complexes. The three major classes of functions associated with them are indicated by various kinds of shading. Of course those of most interest now are the immune response gene regions, labeled I in most species, D and DR in the human HLA complex. The so-called classical histocompatibility antigen regions are now known to be involved in the determination of cell-surface markers for cytotoxic reactions against foreign grafts and against virus-infected target cells. The regions determining complement components are the most recent addition to this set of immune functions associated with these complexes. We'll have more to say about these complement components later.

You'll note that there are some duplications within complexes shown in Figure 1--duplications with respect to the classical histocompatibility regions. This is a common feature of the MHCs. We'll see more examples of that as we proceed. You'll also note that there is a difference in the order of genes in the mouse H-2 complex relative to the other complexes. I might just point out that if you break the H-2 complex between the S and D regions and invert that segment, you'll have the same order as that seen in the other complexes. I think the important thing about these genes is not their specific order, but the fact that they have remained closely associated over rather long periods of evolutionary time.

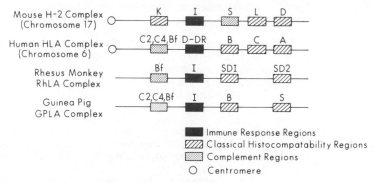

    ■ Immune Response Regions
    ▨ Classical Histocompatability Regions
    ▦ Complement Regions
    ○ Centromere

FIGURE 1

Of course the HLA system is particularly important because of its medical implications, which began with an interest in its role in transplantation reactions. More recently, the HLA system has emerged as a major genetic factor in disease susceptibility. Table 1 shows some examples of the kinds of things that have stimulated a great deal of interest in this human major histocompatibility system. The clearest evidence for a major role in disease susceptibility came with the recognition that patients with ankylosing spondylitis have a high frequency of HLA antigen, B27. Frequencies as high as 90% are seen in patients, but less than 10% in controls, giving a relative risk (meaning the risk for those individuals carrying that particular antigen, relative to the risk for individuals who have some other antigen) of 90. You'll also see in Table 1 a number of other diseases in which such associations have been observed with the D antigens (those associated with the I-equivalent region of the HLA complex), presumably reflecting effects of immune response genes. For example, 100% of patients with nasopharyngeal carcinoma have the marker D Singapore 2--suggesting a very strong association.

Table 1. Some Associations of Human Disease with HLA Type

Disease	HLA Antigen	Frequency in: Patients	Controls	Relative Risk
Ankylosing Spondylitis	B27	> 90%	<10%	90
Coeliac Disease	Dw3	96%	27%	73
Juvenile Diabetes	DRw3	58%	14%	8
Rheumatoid Arthritis	DRw4	70%	28%	6
Nasopharyngeal Carcinoma	D-Sin2	100%	10%	∞

Of course these associations in man were looked for because of an earlier demonstration that H-2 has an important role in immune responsiveness and disease susceptibility in the mouse. The most important finding was the demonstration in 1964 by Lilly that H-2 type determined susceptibility to Gross virus-induced leukemogenesis. That finding opened up a great deal of work and a great deal of interest in the biological roles and significance of the major histocompatibility complexes.

Of course the mouse system has become the most important of the experimental models for functional studies of MHCs because of the availability of inbred strains as constant sources of material for experimental analysis and because of the substantial degree of genetic definition of the H-2 system. There are several reasons for this genetic definition. The availability of inbred strains has made it possible to define the products of genes in the complex · serologically, so it has become possible to define multiple genes within the H-2 complex. Secondly, we owe to George Snell of the Jackson Laboratory the development of H-2-congenic inbred strains. These are strains which have a constant genetic background, for example that of the B10 inbred strain, to which a new H-2 type has been transferred by continuous backcrossing with concurrent selection for a new H-2 type. The strain ultimately derived is thus a B10 strain in all of its chromosomes except for a small chromosomal segment carrying a new H-2 complex, transferred from another strain. This makes it possible to eliminate background genetic variations. To date, more than 150 congenic strains have been developed carrying various discrete H-2 types.

The genetic definition of the H-2 system also reflects the capability to produce and screen large numbers of progeny for recombination and mutation. In our own laboratory, we've produced about 50 H-2 recombinant strains, resulting from more than 13,000 individuals screened. In a number of other laboratories, something like 50,000 mice have been screened for mutations over recent years, with more than 20 H-2 mutants detected and preserved. Those mutants have been particularly useful for functional studies of the genes in which they have occurred.

As a result of the possibility to achieve this genetic definition, a large number of discrete genes within the H-2 complex have been identified and separated from one another, by genetic and biochemical techniques. In Figure 2, 14 genetic loci and 5 regions of the complex are depicted.

SEROGICALLY DEFINED LOCI OF THE H-2 COMPLEX

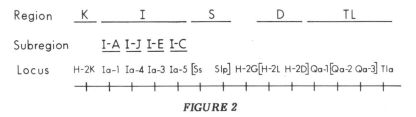

FIGURE 2

Products associated with each of these 14 loci have been separated from one another by recombination analysis and/or by molecular characterization. It is interesting that these are localized to five discrete regions, in each of which there are multiple genes with related functions. We'll see more about that as we proceed. Functionally, the K/D, I and S regions are concerned with cytotoxic reactions, immune response reactions, and complement, respectively. The TL region has become more extensively defined in the past few years. It

106

develops that it determines a series of markers, apparently differentiation markers, on T lymphocytes.

My own interest in the H-2 system grew out of thesis work here at Caltech on a particular serum protein variant. We initially detected this serum protein during a demonstration of immunodiffusion techniques in Ray Owen's immunology course. We found two broad classes of inbred strains with respect to the level of this protein in the serum; some were classified as "low", others "high", on the basis of the strength of the immunodiffusion reaction. The magnitude of the difference is about 20-fold. We called this protein Ss (to denote the "serum substance" variant system). We showed fairly quickly that Ss levels are controlled by a single autosomal dominant gene. Rather early on, we recognized in classifying various inbred strains for the Ss marker that all those strains that had the low Ss level had the same H-2 type, H-2^k, whereas all those that had high Ss levels had other H-2 types than H-2^k, indicating an association with H-2. That led to some linkage tests which demonstrated that there was such close linkage that there was, in fact, no recombination between these two traits.

By that time, Peter Gorer in London had already demonstrated that genes controlling certain serologically-defined markers of the H-2 complex were separable by recombination and could be mapped to at least two regions, K and D. So we became interested in mapping the determinant of Ss variation with respect to those H-2 regions that had been defined by Gorer. To do this, we began screening progeny of H-2-Ss-heterozygous parents for recombination between K and D markers. We initially detected two pairs of reciprocal cross-overs which had the same serological markers in the K and D regions (Figure 3).

RECOMBINATION MAPPING OF THE SERUM
SEROLOGICAL VARIATION

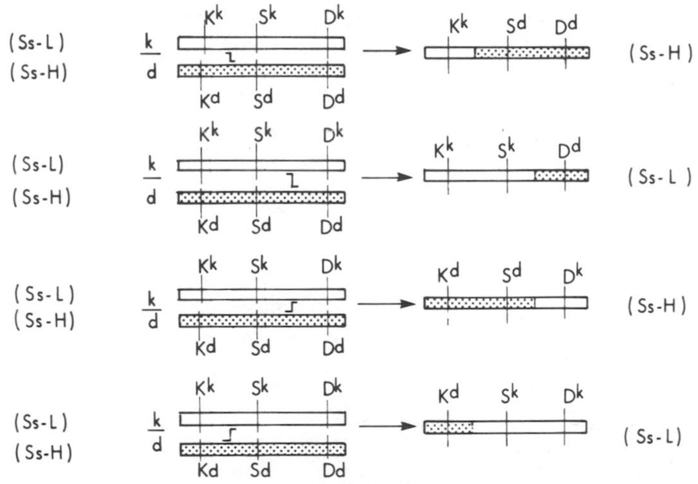

FIGURE 3

One recombinant in each pair had the low Ss type, the other had the high. That indicated that the determinants for Ss variation must map between the separable K and D loci that control serologically-defined H-2 factors, and that defined the S region. The new recombinants were fixed as congenic inbred strains and, because the two strains of each pair are entirely identical, except

107

for the $\underline{S}$ region difference, these have become very useful for functional and biochemical comparisons specifically with respect to that single region.

As we analyzed our new recombinants serologically, it emerged that, in some cases, a particular antigenic marker would seem to be determined by a gene to the right of the $\underline{Ss}$ locus (in the $\underline{D}$ region), in other instances by a gene to the left of the $\underline{Ss}$ locus (in the $\underline{K}$ region). It became impossible to establish a consistent linear map. In time we recognized that one explanation for this difficulty might be that the genetic determinant for this marker was, in fact, associated in some strains with the $\underline{K}$ region and in other strains with the $\underline{D}$ region. The implication of this was that the $\underline{K}$ region product and the $\underline{D}$ region product might be structurally similar. That led us to postulate a "duplication model" for the major histocompatibility complexes. The idea was that, at some point in evolutionary time, in some ancestral species in which there had been a single primordial histocompatibility ($\underline{H}$) gene, some segment of the chromosome carrying that gene underwent a tandem duplication, yielding two $\underline{H}$ genes. Through mutational divergence, these ultimately became the discrete $\underline{H\text{-}2K}$ and $\underline{H\text{-}2D}$ genes of the mouse. Likewise, in the human, their counterparts became $\underline{HLA\text{-}A}$ and $\underline{HLA\text{-}B}$, and so forth in other species. However, it was postulated that the two gene products retained sufficient homologies--structural similarities--to account for the finding of serological similarities.

Biochemical analyses of the products of these two genes eventually established their structural homology. Both genes control a large polypeptide subunit of about 45,000 daltons. Both such polypeptides are associated with a smaller, beta-2 microglobulin subunit. The two kinds of products have a common carbohydrate group. The serological differences in these products are determined by differences in the amino acid sequences of the large polypeptides. Through the work of a number of groups (including Lee Hood's here at Caltech), amino acid sequence homologies of these products have now been demonstrated. Much of the sequence is common to all $\underline{H\text{-}2K}$ and $\underline{H\text{-}2D}$ products, thus demonstrating substantial homology between the products of the distinct $\underline{H\text{-}2K}$ and $\underline{H\text{-}2D}$ genes. Homologies between $\underline{H\text{-}2}$ products and the products of $\underline{HLA}$ have also been shown, indicating that these similar molecules in various species are evolutionary homologs, as well.

DUPLICATED LOCI OF THE $\underline{H\text{-}2}$ COMPLEX

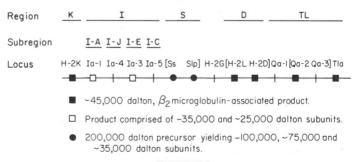

FIGURE 4

In fact there are many duplicated loci within the $\underline{H\text{-}2}$ complex. In Figure 4, the same loci that were represented in Figure 2 are shown, except that we've indicated in Figure 4 those loci that we believe to be duplicate genes in various regions of the complex. The black squares show $\underline{H\text{-}2K}$, $\underline{H\text{-}2D}$ and three other loci whose products all have 45,000 dalton polypeptide chains associated with a beta-2 microglobulin subunit. So the indication is that there has been a great deal of duplication within this complex. The suspicion is that there may be still more duplicate loci whose products will show these same properties. Biochemical analyses of serologically-defined products of genes in the I region have shown that their products are comprised of two polypeptide subunits of 35,000 and 25,000 daltons. There are at least two loci within the I region that produce such products. Products of the two loci have some sequence homology, as shown by peptide maps. Also, current evidence suggests that in the S region there are two loci, each of which determines a 200,000 dalton precursor polypeptide that, on processing, yields three polypeptide subunits, of 100,000, 75,000, and 35,000 daltons. Thus, the MHCs seem to have evolved by multiple gene duplication events and there seems to be a localization of specific classes of duplicate genes to specific regions of the complex.

In 1964 Frank Lilly showed that differences in H-2 types determine differential susceptibility to Gross virus-induced leukemogenesis. Shortly after that, Benacerraf and McDevitt demonstrated MHC-linked immune response genes. Evidence with the mouse in 1968 showed differences among strains in capacity to produce antibodies to some synthetic polypeptide antigens. Different strains showed different patterns of responsiveness to these particular antigens and it was discovered, when H-2-congenic strains were classified for this responsiveness, that the capacity to respond or not to respond was associated with H-2 type. Following that initial demonstration, differences in responsiveness to a large number of distinct antigens were found to be linked to the H-2 complex. Analyses of available H-2-recombinant congenic strains made it possible rather early to demonstrate the existence of a separate H-2 region, called the I region, which was shown to control responses to many antigens. Analyses of responses to certain of these antigens in appropriate H-2-recombinant strains made it possible to further subdivide that I region into at least three separate subregions that control different immune responses. In all cases, the definitive recombinant strains have been preserved and therefore are available to be used for investigations to define the specific functions of the genes carried by those subregions.

In the late 1960's and early 1970's, it became recognized that antibody production involves interactions between the so-called B lymphocytes (bone marrow-derived lymphocytes), that produce antibodies, and the T lymphocytes (thymus-derived cells), that regulate the B cell antibody responses through either "helper" or "suppressor" effects. Analyses of the immune response differences determined by the I region suggested that these immune response genes act at the level of the T cell.

In 1972 I spent some time in Basel, Switzerland, at the Institute for Immunology and participated in a study there with Berenice Kindred in which we found that restoration of antibody responses to athymic mice, by transfer of normal T cells, was H-2-restricted. Dr. Kindred had bred congenic mice of the BALB strain carrying the nude gene, which causes absence of the thymus. These mice are normally unable to give an antibody response. However, when those BALB-nude mice were given thymocytes (T cells) from normal BALB mice, that were identical in all other respects, their capacity to give antibody responses was restored. When they were given thymocytes from other strains, with other genetic backgrounds, but of the same H-2 type, their capacity to respond was also restored, despite multiple non-H-2 genetic differences. On the other hand, if they were given thymocytes from strains with differing H-2 types, there was no restoration. When F1 hybrid thymocytes were given, there was a poor restoration. When a backcross analysis of capacity to restore was carried out, the thymocytes of homozygous, H-2-identical mice restored in 100% of cases, whereas there was a low or intermediate level of restoration when H-2-heterozygous donors were used. That suggested that H-2 compatibility is required for the T cell-B cell interactions that are necessary to promote antibody responses. That initial observation was picked up and studied quite extensively by Katz and Benacerraf, who demonstrated rather clearly that the required compatibility is mediated by genes in the I region. That finding suggested that products of immune response genes might be involved in those T cell-B cell interactions that lead to "helper" effects on antibody production. It was shown later by Tada and others that such I region restrictions may hold for other kinds of T cell regulatory interactions, for example immune suppression.

Although the precise molecular mechanisms for these restrictions were not then and are still not clearly understood, a very striking parallel developed in some studies by Zinkernagel and Doherty, by Shearer, and by others beginning in 1974. These were studies of T cell-mediated cytotoxicity (T cell-mediated killing reactions) against virus-infected cells. The observation made was that, when a "killer" cell of a particular H-2 type has been induced by a virus-infected cell of that same H-2 type, if the virus-infected "target" cell is also of the same H-2 type, there is a killing reaction, whereas, if the virus-infected "target" cell is of a different H-2 type, there is no killing. That showed rather clearly an H-2 restriction on the interactions leading to cytotoxicity, but in this case the restrictions were found to map, not to the I region, as shown earlier, but to either the K or the D region. That implied that there may be rather parallel mechanisms involving, on the one hand T cell-B cell interactions leading to helper or suppressive effects that are mediated by the I region products, and on the other hand interactions leading to cytotoxic reactions that are mediated by K and D region products.

An important approach to the genetic and functional analysis of these effects was the use of mutations in the H-2K and H-2D genes. An example of this is the series of alterations seen as a result of the mutation from H-2K^k to H-2K^{ka}. This apparently was a point mutation in the H-2K gene, since only a single peptide alteration in the 45,000 dalton chain has been shown, in work in Lee Hood's lab. The consequences of that point mutation were a serologically detectable change in the mutant product, skin graft rejection between mutant and parental strains, reactions in the mixed leukocyte reaction, in the graft-versus-host reaction, and in the cell-mediated lympholysis transplantation assays. There is also a restriction of cytotoxic recognition, such that immune cells from a mouse of the mutant type can kill virus-infected target cells of the mutant type but not of the parental type. Such restriction provides evidence that the cytotoxic mechanism is very specifically mediated by the H-2K gene product. Other mutants have demonstrated this for the H-2D product as well.

This kind of mutation analysis has opened up a variety of insights into the potential functions of MHC products. The currently available mutants are all in the H-2K and H-2D genes. Obviously, mutations of genes in other regions of the complex will be of similar importance, if they can be found. The present indication is that there are at least several sets of essentially parallel MHC mechanisms, with I region products involved in those which help or suppress the antibody response by B cells (and the indication is that these two different regulatory mechanisms involve at least three different subregions of the I region), whereas the K and D products are involved in cytotoxic reactions. All of these mechanisms appear to function at the level of T cell recognition of antigen, because of the antigen specificity of all these reactions.

The current interpretation of these results is that the MHC products must function as markers for antigen recognition by T cells and consequent functional triggering of those cells. The indications are that there are a number of different T cell populations that recognize a corresponding number of different MHC products and carry out different regulatory or effector functions when stimulated by antigen. The precise mechanisms of such recognition, for example, whether the effector T cell has two separate receptors, one for an MHC product and one for the antigen, or whether it recognizes some sort of complex of the MHC product with the antigen, are currently the subject of a

110

great deal of investigation by cellular immunologists. The broad significance is that MHC products play a role in cell–cell recognition processes and cell–cell interactions that mediate a variety of regulatory functions involving T lymphocytes.

Given this role in immune regulatory mechanisms, the direct detection, definition and molecular characterization of the I region gene products is obviously a very important matter. It became possible for us to detect and analyze some of these products serologically in 1973, when we completed the breeding of some recombinant congenic inbred strains that are identical in their K and D regions but differ in their I regions. It was possible with the three recombinant strains shown in Figure 5 to define two sets of products, one associated with the I-A subregion, the other with the I-E subregion. (This work was done with Chella David in my laboratory, with a great deal of help from Jeff Frelinger, who had at that time just joined us from Ray Owen's lab.) The products defined serologically were shown to be associated primarily with B lymphocytes, macrophages, and a small population of T lymphocytes. In collaborations with Susan Cullen, it was demonstrated that both of these products have two subunits, with molecular weights of about 32,000 and 25,000 daltons.

H-2 TYPES USED IN ANTI-Ia PRODUCTION

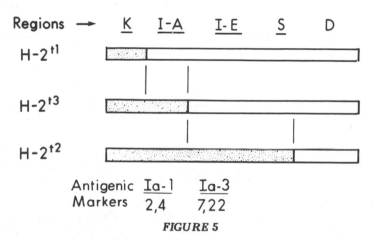

FIGURE 5

Peptide mapping studies have indicated some homology between the products of the I-A and I-E regions. Amino acid sequence analyses have shown homology with DR products. Later, antisera were produced that define specific antigenic markers on suppressor T cells; these markers were mapped to the I-J subregion.

Antisera to I region antigens have also been very useful for analyses of the roles of these products in immune functions. For example, it has been shown that one can, with these antisera in the presence of complement, eliminate certain T cell subpopulations that have suppressive effects and others that have helper effects. B cell subpopulations that produce IgG antibodies and macrophages that present antigen to T cells can likewise be eliminated. One can remove various kinds of soluble effector factors produced by T cells or macrophages on adsorbent columns of anti-I region antibodies. So, these antisera are proving very useful in the dissection of the products and the

111

functions of the genes in the I region and the cells that express them. As a result of that sort of analysis, it has been possible to subdivide the I region into at least four major segments, whose products have different cellular expressions and different kinds of functions. The I-A subregion has products expressed on B cells, macrophages and helper T cell factors. The I-J subregion determines products on suppressor T cells and on a subclass of helper T cells. I-E products are on B cells and macrophages; I-C products on a class of suppressor T cells. The functions associated with these products appear to reflect a series of rather parallel but discrete regulatory mechanisms, all of which perhaps evolved at some point in time from a common ancestral "I" gene.

Now, my story began with the Ss serum protein variation. For many years that was a useful marker for the H-2 complex and facilitated the genetic analysis of the H-2 complex, but we didn't know much about the protein. A few years ago, with Tommy Meo, we recognized that there is immunochemical cross-reactivity between the Ss protein and the human C4 component of complement. More recently, with John Atkinson, we have developed specific assays for C4 activity and have been able to demonstrate that there are quantitative differences in C4 activity that correlate with the differences in level of the Ss protein. Through analyses of appropriate H-2 recombinants, we have been able to show that those differences in C4 activity map to the S region of the H-2 complex, the same location as the Ss gene. All of the current evidence thus suggests that the Ss protein is the mouse equivalent of human C4. The finding of an association of complement activity with the mouse MHC led people working with MHCs of other species to look also for complement associations and three complement factors (C2, C4, Bf) have now been associated with the human HLA system and with the guinea pig GPLA system. It is also significant that there may be differences in complement levels associated with the chicken MHC. If verified, this would imply that the complement components have evolved as a part of the MHC, going back to its early origins--at least to the evolutionary branch point between birds and mammals.

As one example of the emerging potentials for molecular genetic definition of the H-2 system, we recently have been able, with Marleen Roos, to biosynthetically label the Ss protein in in vitro macrophage cultures and to examine the radiolabeled Ss protein found in the culture medium by SDS polyacrylamide gel electrophoresis and peptide mapping. We have also examined a nonfunctional (in the C4 assay) but homologous protein that is also controlled by the S region (called the Slp protein--Slp because it is sex-limited and under testosterone regulation). The interesting thing about the comparison of these two proteins (Figure 6) is that they have compensating molecular weight differences (of about 2000) in two of their polypeptide subunits (called alpha and gamma). We have demonstrated in lysates of macrophages cultured in the presence of radiolabeled amino acids that there are two precursor polypeptides for these two molecules, both with molecular weights of about 200,000. In pulse-chase experiments, after 1 hour of labeling, the precursors decline with time concordantly with the appearance in the medium of the separate alpha, beta, and gamma subunits. Peptide map comparisons, with Keith Parker, of the Ss and Slp molecules show many shared peptides, but also multiple peptide differences, indicating that these are related, but distinct, proteins. We have therefore postulated that there are two structural genes within the S region, Ss and Slp. Each structural gene specifies a single polypeptide precursor with a molecular weight of about 200,000. While the products of these genes show a great deal of structural homology in peptide maps, there must have been a shift in a proteolytic processing site on one of the precursors, to account for the compensating differences in molecular weights of the alpha versus gamma chains. Such approaches promise to give us new insights into the molecular genetics of the complement components. The S region is also of great potential interest for study of mechanisms of regulation of gene expression, since expression of the <u>Slp</u> gene is subject to induction by testosterone.

A MODEL FOR Ss AND Slp SYNTHESIS AND PROCESSING

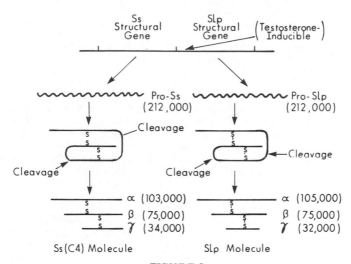

FIGURE 6

Thus, as a consequence of the increasingly detailed molecular definition of <u>K</u>, <u>D</u>, <u>I</u> and <u>S</u> products, in addition to their continued important roles in transplantation, disease susceptibility and immune mechanisms, I think we can look forward in the next few years to significant advances in the application of the kinds of approaches for detailed resolution at the level of DNA in chromosomes that have been discussed in this symposium, and thereby a vastly enhanced understanding of MHC genetic organization and regulation at a molecular level.

Selected Readings

Ross, M. H. J. P. Atkinson and D. C. Shreffler. Molecular characterization of the <u>Ss</u> and <u>Slp</u> (C4) proteins of the mouse <u>H-2</u> complex: Subunit composition, chain size polymorphism, and an intracellular (<u>Pro-Ss</u>) precursor. <u>Journal</u> <u>of</u> <u>Immunology</u> **121**:1106-1115 (1978).

Schwartz, B. D. and D. C. Shreffler. Genetic influences on the immune response. In: <u>Clinical</u> <u>Immunology</u>. C. Parker, ed. W. B. Saunders Co.: New York (1979).

Shreffler, D. C. The <u>S</u> region of the mouse major histocompatibility complex (<u>H-2</u>): Genetic variation and functional role in complement system. <u>Transplantation</u> <u>Reviews</u> **32**:140-167 (1976).

Shreffler, D. C. and C. S. David. The <u>H-2</u> major histocompatibility complex and the <u>I</u> immune response region: Genetic variation, function and organization. <u>Advances</u> <u>in</u> <u>Immunology</u> **20**:125-195 (1975).

SIMPLE SOCIAL CELLS

Dale Kaiser

Department of Biochemistry
Stanford University Medical Center

I would like to tell you about myxobacteria, which are among the simplest of social cells. Myxobacteria have the simple structure characteristic of bacterial cells, yet they behave in complex multicellular ways. First, some examples of their multicellular behavior:

Fruiting

Starvation induces a million or more cells to aggregate together and construct a fruiting body. Within a fruiting body individual cells become spores, which can survive the absence of nutrient for long periods of time. Each genus of myxobacteria has its own pattern of aggregation. The aggregates formed by Archangium, for example, execute large-scale rotary motions in which cells enter and leave the vortices and pile on top of each other. They appear to be trying to select the site on which to construct their fruiting body. Later, vortex motion ceases and the cells enter linear streams that converge on the chosen site. The fruiting body grows into a haystack as cells pile on top of one another. The fruiting bodies of myxobacteria can have surprisingly complex morphology, considering their humble origin. Chondromyces, for example, starts by constructing a haystack of cells like Archangium. But then, a stalk of cells forms beneath the stack and, like a hydraulic lift, hoists the stack into the air. Finally, the stack cleaves into finger-like segments before individual cells within each segment become spores. The final structure resembles a miniature tree about 1/10 of a millimeter tall.

Swarming

Thousands of cells can move together in a myxobacterial swarm, which behaves like a swarm of bees. If one of the spore-containing segments of a Chondromyces fruiting body is placed on nutrient medium, the spores germinate, cells grow out and all the cells assemble in a loose, disc-shaped mass, the swarm. Individual cells within the swarm appear to move independently, and they may leave its edge temporarily, but they quickly return to the group. The whole swarm then moves as a unit. Two separate swarms may fuse to form a single new coherent unit.

When myxobacteria move over a surface, they leave behind them a trail of slime that marks a path for other cells. Other cells, encountering a slime trail, usually turn and follow it. When moving on a slime trail, cells often meet others with whom they join to form little groups of cells that move together along the trail. A slime trail thus serves to bring cells together.

Myxobacteria move only on surfaces; they cannot swim like most other bacteria. Cells move smoothly in the direction of their long axis with occasional reversals of direction. They can move either singly or in groups, but groups predominate. Moving groups are not permanent, however, and frequently break up into smaller units or fuse to form larger ones.

Myxobacteria live in the soil where they lyse and digest other bacteria, algae, and fungi by means of extracellular proteases, lipases and nucleases. Lysis of food cells may be the reason that myxobacteria have evolved multicellular behavior like swarming and fruiting. The idea is that the concentration of lytic enzymes produced by one cell is too low to support rapid growth, but

114

many cells together raise the total enzyme activity and like a cellular wolf pack are able to feed more rapidly and thus grow at a faster rate. Experiments by Dworkin and Rosenberg have shown that cells do grow faster at higher density when they feed on a complex substrate such as the protein casein.

Genetics of myxobacterial motility

Our aim is to understand, in molecular terms, multicellular behavior and how cells interact with each other. The strategy we have adopted is to isolate and investigate nonmotile mutants. If enough nonmotile mutants are isolated, they should reveal most of the genes whose products are necessary for motility. The strategic principle is that of genetic saturation and we decided to start with motility mutants because many of the cooperative ventures of myxobacteria involve the control of movement.

Fortunately, it is rather easy to isolate motility mutants. Colonies formed by nonmotile mutants have a characteristic shape that is easy to pick out. A colony of motile bacteria expands by swarming outward (Figure 1).

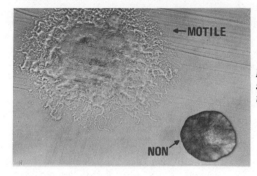

FIGURE 1 — Colonies of a motile strain (top) and a nonmotile mutant (bottom).

Flare-like fingers of cells project outward from the colony edge, and the colony expands, becoming rather flat as cells move away from the center. However, the colony formed by a nonmotile mutant (Figure 1) has a sharp edge and is heaped up because cells cannot escape from the center. Several hundred independent nonmotile mutants were isolated using these morphological differences.

As he was isolating mutants in this way, Jonathan Hodgkin discovered an unexpected thing--that a mixture of two different nonmotile mutants would give rise to moving cells as illustrated in Figure 2. The colony formed by each nonmotile mutant by itself has a sharp fringe-less edge, as in Figure 1, but the mixture has flares of swarming cells. In these flares, some cells are found at least 10 cell lengths away from the mixed-colony edge, but since we don't know from where they started, they may have moved much farther than that. Cells that had moved were picked up, cloned, and examined. They were found to be nonmotile like the starting strains. In other words, no genetic change had occurred in the mixture; rather, there was a transient change in phenotype that led to cell movement. We call this phenomenon "stimulation" to emphasize that it is phenotypic, not genetic. Stimulation requires that cells touch each other, or at least come close enough to touch, for if cultures of the same two mutants are put down on a surface close but not touching, then no cells move out. This doesn't exclude passage of a substance from one cell to another through the medium at very short range, however.

Two mutants that stimulate when mixed must differ from each other. By this criterion, 100 nonmotile mutants isolated from the same motile strain divided into six different types, called A, B, C, D, E, and F. All mixtures of the possible pairs of heterologous types showed stimulation. Mixtures of all pairs of mutants of the same type showed no stimulation. Evidently there is no hierarchy among the types. Nevertheless, there is an important difference between type A and the rest. If both members of a pair are either B, C, D, E, or F--i.e., if both are non-A--then both members move. On the other hand, if the pair is A plus B, C, D, E, or F, then only the B, C, D, E, or F cells in the mixture move. Thus A-type cells never move, but they can stimulate all five other types to move. It is as if there are five different substances needed for movement that can be transferred from one cell to another when cells touch. On this view, a mutant of type B, for example, cannot make "B substance" and so cannot move, but it can receive B substance from any mutant which can make B--an A, C, D, E, or F mutant. If type A mutants were defective in making something that cannot be transferred from one cell to another, the failure of A mutants to stimulation would be explained.

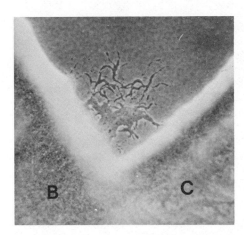

FIGURE 2 — *Stimulation of motility. Suspensions of two different nonmotile mutants, called B and C, were placed on agar so as to partially intersect each other. The edge of the two suspensions forms a large V in the photograph. At the upper left and right where there is no mixing of B and C cells, the edges are sharp showing that B and C are nonmotile. In the center where the two suspensions intersect and mix, a flare-shaped mass of cells that have moved away from the original edge may be seen.*

This view of stimulation is supported by the arrangement of the mutant sites corresponding to the six types of mutants. First, in a random sample of nonmotile mutants, nine-tenths turn out to be type A. Why should most of the mutants belong to one type and only a few mutants to many types? Results of many genetic crosses by transduction between mutants are summarized in the map shown in Figure 3. The A-mutants fall into at least 17 different clusters of sites. Each cluster might be one gene or a cluster of closely linked genes, but for simplicity, I will call each a gene. Evidently there are many type-A mutants because there are many A genes.

In contrast to the multiplicity of A genes, all 17 known sites of mutation to type B are in a single cluster (Figure 3). Moreover, there are no mutants of other types within the B cluster. Similarly, all the C mutants form another cluster, all the D mutants another, the E mutants another, and there is only one F mutant. Thus, there is one-to-one correspondence between stimulation type, as determined by the cell-mixing test, and locus of mutation on the recombination map, as determined by transductional crosses.

Besides stimulation, there is another kind of interaction between cells that is evident in the distribution of cells within a swarm. In the course of isolating and genetically analyzing motility mutants, about 50 different partially motile strains were found. All 50 showed one or the other of two basic cell distributions and we believe that these distribution patterns are two components of which normal motility is composed. One distribution pattern has mostly single isolated cells; the other has cells arranged in groups or islands of about 10 cells with few single cells. Motility genes are associated with one or the other of these two patterns. All of the genes discussed above belong to what we call gene system A, for adventurous, because any strain that has a complete set of A genes can move as single cells. So far 22 genes are known in system A: 5 are stimulatable (genes B, C, D, E, and F) and 17 are nonstimulatable (the 17 clusters of stimulation type A discussed above; please note the distinction between "type A" and "system A"). All but two genes in Figure 3 belong to system A; the exceptions are rif and T.

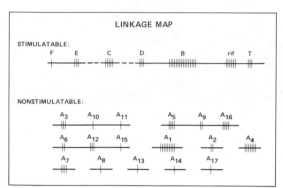

FIGURE 3 — A linkage map showing the clustering of sites that correspond to the same stimulation type.

The rest of the known motility genes belong to system S, which stands for social, so named because swarms formed by strains with a complete set of system S genes have cells arranged in rafts or groups. Cells within a group evidently interact to form the group. The group interaction differs from stimulation in that it occurs among cells all having the same genotype. The difference between the interaction of cells in a group and stimulation is further illustrated by the existence of a stimulatable S-system mutant, called type T, which maps near the stimulatable A-system mutants. There are also eight different nonstimulatable S-system genes, but the genes of system S are not yet saturated with mutations and new genes are still being found. Wild-type myxococcus employs both systems for its movement. The two systems are genetically almost completely independent of each other, as they have only one gene in common. Apart from mutants in the common gene, all other nonmotile mutants are double mutants with one mutation in a gene of system A and another in a gene of system S.

Part of the behavioral difference between socially moving cells and adventurously moving cells arises from the presence of pili, long thin hairs that project from the cells' surface. Many kinds of bacteria have pili, but myxobacterial pili are unusual in that they emerge only from ends of cells, never from the sides. The production of pili is under the control of genes that belong to system S, as is evident from an examination of different types of strains. Wild-type strains have all of the genes of the A-system and all of the genes of the S-system. Wild-type swarms have both single cells and groups of cells, showing a composite of the A and S patterns. Wild-type cells bear pili.

With two systems, A and S, four genetic states are possible, wild-type (A^+S^+) and three others. Colonies of strains representing each of these states are shown in Figure 4 along with a symbol to indicate whether pili are found to be present or absent.

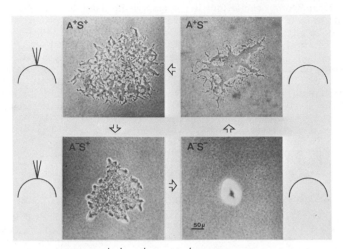

FIGURE 4 — *Colonies of A^+S^+, A^+S^-, A^-S^+, A^-S^- showing how the distribution of cells at the colony edge differentiate these four genetic states. Diagrams to the left and right show a rounded end of a cell and indicate for each state whether the hair-like pili are present (a bunch of 3 lines) or absent (the ends are bald).*

Mutants that are defective in one or another gene of system A, but have a complete system S are in a state symbolized A^-S^+, which is partially motile. The edge of the swarm of an A^-S^+ colony has only groups of cells and such cells also have pili. A nonmotile mutant has a defect in any gene of system A and a defect in any gene of system S. Such a double mutant, A^-S^-, forms a colony that has a smooth edge, and the cells are bald; they lack pili. The A^+S^- state was generated by transducing an A^-S^- strain with phage grown on an A^+ donor. The edges of colonies of these strains have typical adventurous swarming-- many single cells. As for pili, these strains are also bald. Finally, to complete the set of states, a complete system S was restored to A^+S^- by transduction to give A^+S^+ again. These cells regain the capacity to make pili. Therefore, pili are controlled by the state of system S and system A seems to have no effect on their production.

Despite the differences in behavior of cells that have only an active A or only an active S system, the two systems share the capacity to be stimulated. But, when the A system is stimulated, cells move with the typical A pattern; likewise, when the S system is stimulated, they move with the S pattern. Evidently, stimulation does not require pili because two different nonmotile, nonpiliated A^-S^- strains can stimulate each other.

Summing up

The apparently complex maneuvers of myxobacterial cells evident in swarming and fruiting involves the simultaneous action of two motility systems--one that gives single-cell movement and another that leads to formation and movement of groups of cells. Two kinds of social interactions have been found. (1) Pili connect cells together and in some way permit groups to form and to move in a way that single cells cannot. A mutant that is defective in single-cell movement (an A^-S^+ mutant) cannot move as individual cells but can move at higher cell density in the form of groups. It's as if a cohort of one-legged men can walk by tying themselves together. (2) Stimulation reveals another kind of interaction between cells. Two different nonmotile mutants, when they come in contact with each other, are subsequently able to move for a short period of time.

Selected Readings

Hodgkin, J. and D. Kaiser. Cell-to-cell stimulation of movement in nonmotile mutants of Myxococcus. Proceedings of the National Academy of Sciences, Washington, 74:2938 (1977).

Hodgkin, J. and D. Kaiser. Genetics of gliding motility in Myxococcus xanthus (Myxobacterales): Genes controlling movement of single cells. Molecular and General Genetics 171:167 (1979).

Hodgkin, J. and D. Kaiser. Genetics of gliding motility in Myxococcus xanthus (Myxobacterales): Two gene systems control movement. Molecular and General Genetics 171:177 (1979).

Kaiser, D., C. Manoil and M. Dworkin. Myxobacteria: Cell interactions, genetics, and development Annual Review of Microbiology 33:595 (1979).

Rosenberg, E., K. H. Keller and M. Dworkin. Cell density-dependent growth of Myxococcus xanthus on Casein. Journal of Bacteriology 129:770 (1977).

SESSION V: NEURONS AND BEHAVIOR

Introduction

Seymour Benzer

James G. Boswell Professor of Neuroscience, Caltech

The first paper of this session is by someone I met when I came to work in Roger Sperry's laboratory at Caltech in 1965, to learn about behavior. I found all kinds of animals, and people working on them. The animals included chicks, frogs, toads, fish, cats and monkeys. Among the people, there were Roger himself, of course, and Jerre Levy, a graduate student. Jerre's presence was very much felt, providing many examples of the glorious complexity of human behavior. The problem of how much of it is inherited has been of continuing interest to both of us. Jerre has had the courage to tackle such questions at the human level, and has done interesting work on the inheritance of handedness in humans, in addition to becoming one of Sperry's star disciples on the hemispheric specialization of brain function. She is now at the University of Chicago. I am happy to welcome back one of our favorite daughters of Caltech--Jerre Levy.

The second talk will be by Jim Gould, who precociously underwent the transition from phage to neurobiology while still an undergraduate at Caltech. His research in Bob Sinsheimer's lab and his later work on honeybees earned him the Undergraduate Research Award of the 1970 graduating class. My course on behavioral biology included a series of lectures on honeybees and Wenner's attack on von Frisch's notion that the waggle dance serves as a means of communication of direction and distance information. Three students in the class--Gould, McLeod, and Henery--decided to do some experiments to resolve the issue. They got a research grant from the undergraduate student organization, mapped out a desolate area in eastern Oregon that was as devoid as possible of any natural landmarks, and set up shop in a tent. From their weekly reports during that summer, one could see the tension rising. The bees gave problems and the researchers were living in cramped quarters without a shower, eating cans of beans. They were almost at each other's throats when the experiments began to work, confirming von Frisch's idea. The three undergraduates published the results as a lead article in Science. At Rockefeller, as a graduate student, Jim looked at other animals, including expeditions to South America to listen to whales' love songs, but in the end he has stuck with honeybees and is deep in research at Princeton, which he will tell us about.

The next speaker is from the outstanding Harvard Neurobiology Department, the source of three recent additions to our own Biology faculty. Ed Furshpan got his PhD at Caltech in 1955 with Wiersma but, until now, he has been an exception to Bonner's Law of Reappearance: this is the first time he has been back to Caltech in 23 years. Now that the ice has been broken, we hope that he will come more frequently.

Ed was actually a pioneer in Drosophila physiology but somehow lost the golden thread and drifted to other organisms. He has for many years collaborated fruitfully with David Potter, to the point that their names are so closely associated that people save time by referring to them as "Furshpot". Their contributions include the discovery of the electrical synapse and the analysis of neural systems in culture. Their entry into tissue culture took place during a summer at the Salk Institute, when several of us decided to compete on the development of a culture system in which one could observe the formation of neuromuscular synapses. The three teams were Steve Kuffler and Monroe

121

Cohen, who chose the frog, Ed Lennox and I, who chose <u>Drosophila</u>, and Furshpot, who chose the mouse. By the end of the summer, the frog people had only gotten up to the level of Harrison's experiments of 1914, and Lennox and I barely reached the point recently achieved by Seecof, and we dropped the project. Furshpot did not get very far that summer, but they stuck with the problem and have continued with it ever since. We look forward to hearing some of their recent results.

Gunther Stent and I overlapped as postdoctoral fellows in Delbrück's lab at Caltech in 1949 and also in Lwoff's lab at the Institut Pasteur in 1951. Gunther originally studied physical chemistry. His transition to biology was stimulated, as was mine, by Delbrück's suggestion of the possible existence of a complementarity principle in biology, i.e., if one were to do all the measurements on a cell necessary to predict its future, that would automatically kill the cell. In other words, complete predictability would be complementary to the living state.

Gunther has always had a philosophical bent and an elegant literary style. At Berkeley in the 60's, he got caught up in the spirit of the times and wrote a book entitled <u>The Coming of the Golden Age</u>, the golden age being the time when we could dance on the grass and let computers do all the work. The book pointed out that progressive abandonment of rules and restrictions was gradually leading to the death of all the arts and sciences (with the possible exception of the movies). Gunther then turned around and contradicted himself by getting tremendously enthusiastic and excited about neurobiology. He took a sabbatical at Harvard, where he was influenced by John Nicholls to work on the leech. The leech has rather large neurons and a simple sinusoidal swimming movement and Gunther and his colleagues spent the next years back at Berkeley working out completely the circuit diagram for this swimming oscillator. Unfortunately, the leech has no genes, i.e., that you can easily manipulate and do experiments with. This, I suspect, has influenced Gunther to argue about the uselessness of genetics in neurobiology. But we all know that this will change the moment he discovers his first mutant leech. In the meantime, he is gradually moving in the direction down toward the gene, and will speak on the problem of where the leech nervous system comes from.

VARIATIONS IN HUMAN BRAIN ORGANIZATION

Jerre Levy

Department of Behavioral Sciences
University of Chicago

After listening to the evidence, presented at these meetings, of the remarkable revolution in biology that has occurred over the last 50 years, I began to consider the period when I was at Caltech looking at people and their brains and Bob Edgar was putting phages together. The relationship between my research activities and the biological revolution was not immediately apparent, but, upon further thought, I discovered the connection. I discovered that our research provided the answer as to why the revolution occurred: in brief, we found that, during the last 50 years, the human brain had evolved, doubling in size from one cerebral hemisphere to two--and naturally, therefore, greatly increasing the intelligence of biologists. It is the story of this amazingly rapid evolution of the human brain that I wish to recount.

Prior to 1836 there was very little evidence that the human brain differed in any significant way from that of other primates except for a certain size advantage. There were no obvious qualitative distinctions that could account for what we assumed to be a cognitive superiority. In 1836, Marc Dax, a French neurologist, presented a paper at a small conference in Montellier, reporting that only the left half of the human brain had speech, a paper that remained unpublished at the time and was little noted. Not until 1861 did the neurological community come to recognize that the human cerebral hemi-spheres were functionally asymmetric when Paul Broca, another French neurologist, published his observations on the association of linguistic disorders with left hemispheric damage.

Subsequent to Broca's report, a great many papers were published confirming the unilateral localization of speech and other aspects of language to the left side of the brain. The left cerebral hemisphere came to be considered the dominant organ of thought, not only for language, but for all sensory understanding, for the planning and execution of behavior, and for all conceptual integration. The right side of the brain became merely an input/output unit, sending information to its dominant partner on the left from the left half of space and relaying commands to the left side of the body. By the end of the 19th century, man had become a half-brained species whose single, thinking hemisphere barely retained even a size advantage over the brain of apes. So strongly was this belief held that when evidence started to emerge that damage to the right hemisphere led to cognitive disorders not observed following left-side damage, the obvious interpretation was ignored. Indeed, many leading neurologists maintained that deficiency syndromes resulting from right hemisphere damage offered the final proof of the left hemisphere's dominance! The cognitive disorders, it was said, were due to a release of pathological neural activity in the left hemisphere in consequence of its loss of inhibition by the right.

Until the early 1960's neurological science was at an impasse: any disorder concomitant with left-side damage was attributed to the left hemisphere's specialization for the disordered function, while any disorder concomitant with right-side damage was attributed to pathological release of activity in the left hemisphere that interfered with its specialized function. Obviously, given the interpretative premises, no collection of observations from patients with unilateral lesions could overthrow this left-sided view of the human brain. The fact that patients with damage to the right hemisphere manifested specific disabilities in recognizing faces, in understanding maps, in rendering three-dimensional relationships in drawings, in copying designs with colored blocks, in understanding fragmented pictures, and in a variety of other problems requiring literal, imagistic memory or the extraction of invariants in spatial relationships, had almost no effect on the widely held conclusion that only the left side of the brain was competent for cognitive processing. Hughlings Jackson, in the 19th century, had suggested that the right hemisphere was specialized for certain perceptual functions, and a few more modern researchers had been timidly suggesting the same thing, but these represented a small and ignored minority.

During the 1940's a group of patients had been available for study who, at least in principle, could have provided a definitive answer regarding the roles of the two hemispheres in psychological function. These patients were epileptics in whom all the neocortical commissures connecting the two sides of the brain had been surgically sectioned in order to alleviate seizures. With the hemispheres disconnected, it would be expected that activities in the two sides would be functionally separate and could be investigated independently. Remarkably, however, no signs of functional separation were seen, leading some neuropsychologists to mysticism and Karl Lashley to the suggestion that the sole function of the commissures must be to support the hemispheres--a suggestion, of course, that was made in jest and with a sense of frustration. As we now know, the failure to observe a functional separation following splitting of the brain was due to an inadequacy of behavioral assessment techniques.

Beginning in 1961, Drs. J. E. Bogen and P. J. Vogel of Los Angeles performed split-brain surgery on another series of epileptic patients, and since that time these patients have been intensively investigated by Professor Roger Sperry and his colleagues and students at Caltech. In the initial report of the first patient in the Bogen and Vogel series, Michael Gazzaniga, Dr. Bogen, and Professor Sperry described a very distinct and dramatic split-brain syndrome: information presented to the right-half sensory field (and projecting to the left hemisphere) could be verbally described, while information presented to the left-half sensory field (and projecting to the right hemisphere) could not. In spite of the expressive aphasia of the right side of the brain, however, the patient could indicate by nonverbal means that his right hemisphere was aware of the sensory input. Further studies showed that there was a complete mental separation between the two hemispheres; there were two sets of perceptions, two sets of motivations, and, in fact, two minds, one in each hemisphere that was completely out of conscious contact with the other. In contrast to these observations in the laboratory, the performance of split-brain patients in everyday life was remarkably normal.

The integrated behavior of these patients outside the laboratory, the lack of any serious dysfunction, indicates that each separated half-brain is a whole and intact brain within itself whose functions are not disrupted by disconnection from the other side. The right hemisphere gives every appearance of being a thinking, behaving, competent human brain, albeit one that cannot speak. It is a brain that understands what it sees, hears, or feels, that has concepts, that can plan behavior, and that can experience emotion. These conclusions were apparent from the earliest studies, and they not only totally overthrew the earlier, left-sided view of man, but they offered the possibility of comparing the two sides of the brain for a variety of cognitive functions. The fact that the right hemisphere was demonstrated to be as competent as the left in understanding simple, nonverbal sensory input and in guiding nonverbal behavioral responses, though raising it considerably above the relay-station status it previously had, was insufficient to establish that it had any special properties of its own, comparable with the special verbal properties of the left hemisphere.

A number of further studies were devoted to this issue, and I should now like to describe a particular set that were conducted in collaboration with

Roger Sperry and Colwyn Trevarthen. The vertebrate visual system is such that if an animal fixates on any given point in space, everything to the left of fixation projects to the right half-brain and everything to the right of fixation projects to the left half-brain. It is, therefore, possible to transmit visual information selectively to one or the other cerebral hemisphere by having human subjects fixate on a specified point while visual stimuli are displayed to the right or left. Since voluntary eye movements require about 200 msec, a display flashed at a shorter duration prevents refixations that would project the stimulus to the unintended hemisphere. A further control for eye fixation may be obtained by recording electrooculograms that indicate the position of the eyes.

In our studies, patients were asked to fixate on a designated point and electrooculograms were simultaneously recorded. During the fixation period, stimuli were flashed for 150 msec or less and patients were asked to respond to these stimuli in various ways. By such means, we could determine which hemisphere tended to dominate the behavioral response, the nature of the response it made, and the adequacy of its processing of the information. Figure 1 shows one of the first tests we gave. Stimuli were made by cutting photographs of faces in half and recomposing them into facial composites, with the half-face of one individual on the left and the half-face of another individual on the right. A composite was flashed as the patient was on fixation, with the midline of the photograph aligned with the center of the visual field, projecting one half-face to the left hemisphere and the other half-face to the right hemisphere.

Though normal people immediately perceive that two different half-faces are being displayed, split-brain patients give no indication of such awareness. Each hemisphere effects a perceptual completion of the half-face it sees and behaves as if it has seen a whole and complete face. Under one condition of the test, photographs of the whole faces were displayed in free vision, patients were told they would see a face briefly exposed in the "machine" (a tachistoscope), and that they were to point to the face in free vision that matched the briefly exposed face. In spite of the fact that each hemisphere was exposed to a half-face, different from that seen by the other side, and regardless of whether patients pointed to their choice with the left or right hand, in the overwhelming number of cases for all patients tested, a single matching choice was selected and the face chosen was that seen by the right hemisphere. The patients behaved as if the left hemisphere had seen nothing at all. Once the selection was made, a selection, moreover, that could be observed by the left hemisphere, patients remained placidly waiting for the next trial. Neither by facial expression nor by verbal response, did the left hemisphere indicate that it disagreed with the choice made. The dominance of the right hemisphere in face recognition was consistent with observations that patients with injury to the right side of the brain often had difficulty recognizing faces. Nevertheless, we found it surprising that in our competitive perception tests, the left hemisphere should behave as if it were unconscious.

We next removed the whole faces from free vision and instructed the patients that they were to describe the faces flashed or, in some cases, to name the face when names had previously been taught. Since the right hemisphere, we believed, was unable to speak, this procedure was designed to force a response from the left hemisphere. With verbal responses required, the left hemisphere typically described the face it had seen, and the right hemisphere, though hearing the description, gave no indication that it disagreed with the response. (The right hemisphere can derive meaning from a great deal of what is said.) We were thus faced with a situation in which patients behaved as if only the right hemisphere could perceive faces when a matching response was given, and as if only the left hemisphere could perceive faces when a verbal response was given.

125

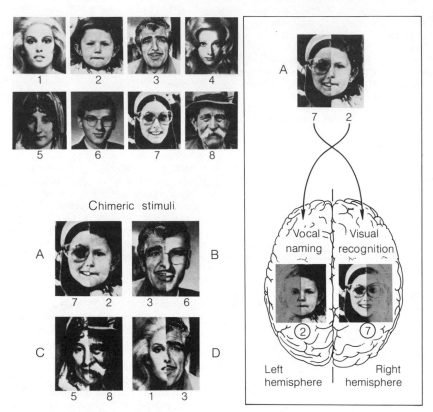

Chimeric stimuli

FIGURE 1 — *Face recognition test administered to split-brain patients. Under one condition, patients were asked to choose one of the complete faces as matching the face they had been briefly shown; under the other condition, they were asked to describe or name the face. [From J. Levy, C. Trevarthen and R. W. Sperry,* Brain **95**, *61-95 (1972). Reprinted by permission.]*

The simplest interpretation of these results was that the nature of task instructions selectively activated one or the other hemisphere, preparing it for sensory processing and for a behavioral response, leaving the other side of the brain unaroused and inattentive. Sperry had suggested in 1952 that "perception is the preparation to respond." By this, he meant that a stimulus is perceived when the resultant spatio-temporal pattern of neural activity has become adjusted for the regulation of a behavioral response. This conception is based on the fact that the function of brains is not the manufacture of ideas, sensations, images, feelings, or the storage of memories, as most laymen suppose, but is, rather, to prepare the organism for adaptive action. Adequate stimulation of sensory receptors and the transmission of neural impulses to their projection regions would be, under this view, insufficient to give rise to a perception. The behavior of split-brain patients under the conditions of our competitive perception tests provides strong support for Sperry's suggestion.

In order to further investigate how the two hemispheres dealt with conflicting information, we gave a series of interrupted trials in which, under one condition, patients were instructed to select a matching face, but on some trials were interrupted before they could respond and told to describe the face, while, under the other condition, they were told to describe the faces, but were interrupted prior to the response and told to select a matching face. In such cases, the patients selected faces matching what the right hemisphere had seen and described faces seen by the left hemisphere. These observations show that information received by the non-attending hemisphere had been available in short-term memory and inaccessible to consciousness until the interruption itself served to activate further processing to the level of awareness, an inference concordant with a phenomenon psychologists have termed the "what-did-you-say effect." When people are engaged in conversation and a third party interrupts the conversation with a remark directed to one of the participants, very often the participant will have very little awareness of what was said and will ask, "What did you say?" However, before the question can be answered, the questioner will suddenly realize what was said. This common occurrence has been interpreted to mean that information is stored below the level of consciousness and reaches awareness only when attention is directed to it. It appears that, in split-brain patients, the same type of phenomenon occurred when a patient's attention was directed to information held by the previously inattentive hemisphere.

When overtly discrepant responses were made in these interrupted trials, patients manifested all the symptoms of psychological defense that one might expect. Some would break eye contact with the investigator; others would attempt to distract the investigator with irrelevant remarks; some would giggle. When the discrepancy was pointed out and an explanation requested, responses ranged from outright denial that any discrepancy occurred ("I did not point to that face!") to casual comments that attention had been wandering. Of course, the verbal denials were, in a very real sense, true, since the left, speaking hemisphere had not, in fact, pointed to a face. Similarly, it was also true that the left hemisphere had been inattentive and not in control of behavior when matching faces were selected.

A variety of other competitive perception tests was also given, in which patients had to match or describe nonsense shapes, common objects, three-element arrays of X's and squares, or words. In all these cases, the right hemisphere strongly dominated the matching response, and its accuracy for matching faces and nonsense shapes exceeded the accuracy of the left hemisphere for verbally labeling these stimuli. However, in matching pictures of common objects or in matching of words, the right hemisphere was no more accurate than the left at naming--both hemispheres were essentially 100% correct at their respective tasks. Further, the right hemisphere was less accurate at matching the three-element arrays than was the left hemisphere at describing them, in spite of the fact that it dominated behavioral control during the matching task. As later tests showed, a hemisphere becomes activated and dominates processing and behavior in accordance with what it thinks it can do, rather than in accordance with its actual relative competency for the particular problem at hand. Evidently, the right hemisphere believes itself to be especially able at detecting visual identities, though this is clearly not true for stimuli whose components have only an arbitrary relationship. The errors of the right hemisphere in matching three-element arrays often consisted of transformations of X's to squares and squares to X's, patterns such as X-X-square being matched to square-square-X, the relation, "same-same-different," being preserved, but the elements composing the relation being lost.

In a certain sense, we were surprised to find a right hemisphere dominance at matching words, since we had assumed these would be encoded according to meaning, and, if so, by the left hemisphere. However, the task itself did not actually require such semantic decoding and words could have been treated as nonsense shapes. In order to test this, we repeated the word task, but patients were required to point to a picture named by the word. This requirement necessitated reading the words for meaning. In contrast to the earlier results, a strong left hemisphere dominance emerged, the picture selected representing the word seen by the left half of the brain. This dominance was not due to incompetence of the right hemisphere at simple reading since when words were projected solely in the left visual field, the left hemisphere seeing nothing, patients could correctly select matching pictures. Rather, the competitive dominance of the left hemisphere at reading must be interpreted to reflect its stronger propensity to assume control. As mentioned previously, the right hemisphere can understand a great deal of what is said and can read at a simple level; though it is deficient at understanding abstract concepts, its linguistic limitations do not appear to lie predominantly in the semantic domain.

Figure 2 illustrates a test designed to assess the right hemisphere's phonetic capacities. Here, composite pictures, as before, were flashed to the two hemispheres, but patients were required to select another picture having a rhyming name (eye-pie, rose-toes, bee-key). Not only was the left hemisphere strongly dominant, but when pictures were shown only to the right hemisphere, it performed at chance. Further, though the right hemisphere could indicate by a nod or head-shake, whether the picture it was shown was one named by the investigator or not, it was totally at chance when asked to indicate whether an object named by the investigator rhymed with the name of the object seen.

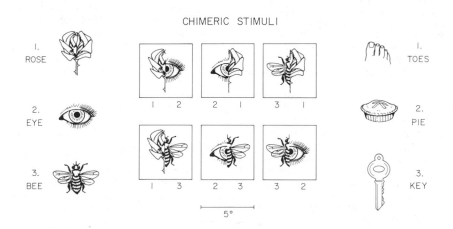

FIGURE 2 — *Rhyming objects test administered to split-brain patients. Patients were asked to select one of the choice pictures having a name that rhymed with the name of the picture briefly flashed.* [*From J. Levy and C. Trevarthen,* Brain *100, 105-118 (1977). Reprinted by permission.*]

Apparently, the right hemisphere's capacity to understand spoken language does not depend on a phonetic analysis of the acoustic stream. The right hemisphere's inability to speak may derive from the fact that, though it can decode sound to meaning, it may not be able to generate the sound image of a concept it possesses, an image that may be necessary for articulatory programming.

In a final test in this series, patients were instructed to select pictures that were either <u>visually</u> similar to the perceived stimulus (a cake on a cake plate and a hat with a brim) or that had some <u>functional association</u> with the perceived stimulus (a cake and a knife and fork). In general, the visual-similarity instructions elicited right hemisphere control, while the functional-association instructions elicited left hemisphere control, and this occurred in spite of the fact that on some occasions, the controlling hemisphere failed to follow instructions, giving visual matches to functional instructions or vice versa. Thus, even when a hemisphere failed to follow instructions and adopted a matching strategy inappropriate to its own specializations, the instructions themselves were effective in selectively activating the hemisphere specialized for the strategy demanded. The hemispheres are evidently specialized not only in terms of actual ability differences, but also in terms of their propensities to dominate processing and behavior in accordance with perceived task demands.

These, and a large number of other studies, suggest that the left hemisphere surpasses the right in its understanding of logical and temporal causality, in its capacity for symbolic representations bearing no isomorphic relations with the concepts represented, and in analytic and propositional reasoning. These qualities or strategies of thought are, of course, associated with its superiority and dominance for linguistic processes. The right hemisphere, in contrast, appears to surpass the left in the capacity for literal representations of its experiences that preserve the rich sensory qualities of the original experience, in extracting invariants from spatial and physical relationships, and in performing spatial transformations on its memorial representations.

The issue I would now like to address concerns the consequences of cerebral asymmetry of function in normal people. Were the pattern of hemispheric lateralization a species-specific characteristic, as invariant in its manifestations as is our bipedalism, there would be little hope of doing more than offering speculations concerning its evolution and adaptive function. One could hardly determine the nature of a covariation in neurological and psychological organizations were there no variance in the former dimension. However, cerebral asymmetry has been found to take many forms: Though the large majority of right-handers have verbal functions and associated cognitive traits predominantly integrated by the left hemisphere and nonverbal functions integrated by the right--as is also the case for a substantial fraction of left-handers--a minority of people have a mirror-reversed organization, and there is considerable variation in the extent of both functional and anatomical asymmetry. Cerebral laterality patterns differ as a function of handedness, sex, and other organismic variables, and to a large extent, these variations appear to be of genetic origin.

The first major deviation from the typical organizational pattern was noted in left-handers, a group that is sinister in more ways than one. First, left-handers are highly variable in the degree of left-hand dominance they manifest, some being almost totally left-handed, others differing from right-handers on only one manual activity. Secondly, left-handedness can be a normal genetic variation or can be a pathological condition resulting from minor damage to the left hemisphere during the prenatal period. Thirdly, even in left-handers who are nonpathological cases, there are enormous variations in the laterality pattern, some having left-side language, some having right-side language, some having partial bilateralization of verbal functions, some having partial bilateralization of nonverbal functions, and some having partial bilateralization of both sets of functions. Left-handers are confusing to neurologists, gauche to the French, and sinister to a large number of cultures.

The heterogeneity of sinistral brain organization has led some current researchers to suggest that <u>all</u> left-handers are brain damaged, that left-handedness itself is a pathological sign. If true, this would be quite tragic

129

considering the contributions that <u>might</u> have been made by Michelangelo or Leonardo de Vinci had they not been brain damaged. However, the brain-damage hypothesis is almost certainly wrong and it is highly probable that the brains of the world's left-handed geniuses were fully intact. The clearest evidence that handedness is, at least in part, genetic comes from studies of neonatal asymmetries in behavior. The majority of infants rotate toward the right during delivery, display a rightward tonic neck reflex during the neonatal period, lead off with the right foot when the stepping reflex is elicited, grasp objects for a longer period with the right than left hand by two months of age, and orient more readily to sensory stimulation on the right than on the left, with a minority of infants displaying the opposite asymmetries. The infant asymmetries are correlated with handedness at ages 2 and 10, and, most importantly, with parental handedness, babies with one or two left-handed parents either showing no overall lateral bias or a leftward bias, in contrast to those having two dextral parents who typically manifest a rightward bias. Further, certain dermatoglyphic asymmetries of the hands are correlated with handedness, and dermatoglyphic patterns are highly heritable and formed by the 18th gestational week.

Though the infant brain is highly plastic in terms of the functions each hemisphere can come to serve if the other side is damaged, there are manifest asymmetries in both function and structure at birth. EEG studies show that the infant left hemisphere is selectively activated by verbal stimuli, while the right is selectively activated by music. The evoked response is larger over the left hemisphere to verbal stimulation and over the right to musical stimulation. By a few months of age, babies have a right ear superiority in discriminating consonant sounds and a left ear superiority in discriminating musical elements. While the majority of babies show the above pattern of lateralization, a minority show the reverse, and a comparable variation is also observed in neuroanatomical asymmetries.

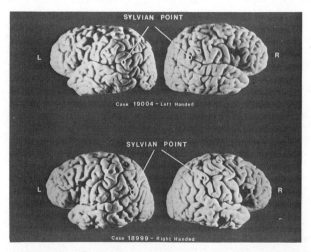

FIGURE 3 — *Lateral surfaces of the cerebral hemispheres of a left-hander and a right-hander. Note the asymmetry in length and placement of the Sylvian Fissures in the dextral brain, and the relative symmetry in the sinistral brain.*

Figure 3 depicts one of the anatomical asymmetries distinguishing left- and right-handed adults. In the typical right-hander, the Sylvian Fissure, separating the temporal lobe below from the frontal and parietal lobes above, is longer and more horizontal in the left hemisphere than right. The areas bordering the Sylvian Fissure are critical for language function and the anatomical asymmetry observed is assumed to reflect the greater development of these areas in the left hemisphere of right-handers. In left-handers, the Sylvian Fissures on the two sides of the brain tend to be more symmetric in length and placement, probably reflecting the partial bilateralization of linguistic processes in a substantial proportion of sinistrals. There are a number of other asymmetries, also, regularly observed in the dextral brain and often reversed or reduced in magnitude in the sinistral brain, all of which have obvious associations with the known functional differences in the two sides of the brain, and with differences in laterality patterns in the two handedness groups.

The most reasonable conclusion from the available evidence is that there is a genetic polymorphism in patterns of cerebral lateralization and their behavioral correlates, but there are several questions raised by such a conclusion. First, we have to consider why the human brain is asymmetric in the first place. In the phyletic series representative of human evolution, the first indication of cerebral asymmetry appears in the apes. As in man, the left Sylvian Fissure is longer in the left hemisphere than in the right, but in monkeys the fissures are symmetric. Further, functional studies of split-brain monkeys conducted by Chuck Hamilton reveal no evidence of lateral differentiation. One is led to the conclusion that asymmetry in the primate line is a rather recent evolutionary development and one that was probably subject to strong selective pressures. An obvious benefit of lateralization would be an increase in cognitive power concomitant with the deduplication of function entailed. A perfectly symmetric brain would be a perfectly redundant brain with respect to higher cognitive functions, and for animals whose fitness was strongly dependent on intelligence, lateral specialization of cerebral function would be expected to confer a significant advantage.

If the above inference is correct, however, it would seem to imply that left-handers ought not to exist. All extant members of our species should be equally and maximally lateralized. Further, no explanation is apparent regarding the cause of variation in the direction of asymmetry. Why do some 6% or so of people have a mirror-reversed organization? Finally, even if some justification can be found for the existence of left-handers, why are they not all simply mirror images of right-handers? How is it possible for a left-hander to write with the left hand if language is predominantly organized in the left hemisphere? Is linguistic information necessary for writing transmitted to the right hemisphere, and, if so, why should such a complex control system exist? Do such left-handers rely on the direct, uncrossed motor tracts for control of the left hand, and, if so, why, since only 20% of the motor fibers run in the uncrossed tract?

In an attempt to gain some clue regarding the answers to these questions, I, in collaboration with a student, Marylou Reid, investigated variations in cerebral laterality in a group of 25 right-handers and 48 left-handers. In addition to sex, subjects were also classified by the hand posture adopted during writing. The large majority of right-handers hold their hands below the line of writing, with the tip of the pencil pointed toward the top of the page--a posture also adopted by perhaps half of left-handers. A small minority of right-handers and about half of left-handers use an inverted posture, with the hand held either parallel with or above the line of writing, with the tip of the pencil directed toward the side or bottom of the page. Similar variations are seen in writing posture in association with handedness in Israel, where writing proceeds from right to left. Given that some American right-handers use the inverted posture, as well as a substantial fraction of Israeli left-handers, it seemed unlikely that the hand posture variation could be a peripheral adaptation to writing with the left hand in a left-to-right direction, and the possibility that it reflected aspects of central brain organization could not be discounted.

Two tests of cerebral laterality were administered, one designed to assess the specializations of the verbal hemisphere and one designed to assess the specializations of the nonverbal hemisphere. On the verbal test, three-letter nonsense syllables, vertically oriented, were flashed to the left or right visual half-field, and subjects were required to pronounce the syllable perceived. On the nonverbal test, a dot was flashed in one of 20 possible positions in the left or right visual half-field, with subjects required to identify the relative location from a dot array displayed in free vision.

Though normal individuals can transmit information received by one cerebral hemisphere to the other via the cerebral commissures, a certain amount of information is lost in the transmission and information processing is superior for material projected directly to the specialized half-brain. Of the 25 right-handers tested, 24 used the normal hand posture and 1 the inverted hand posture (Group RN and Subject RI, respectively). Group RN showed, as expected, a strong right visual field (left hemisphere) superiority on the syllable test, and a strong left visual field (right hemisphere) superiority on the dot location test. Subject RI displayed the opposite asymmetries on the two tests. Of the 48 left-handers, 24 used the normal posture (Group LN) and 24 used the inverted posture (Group LI). Subjects in Group LN were completely mirror-reversed from those in Group RN, having a right hemisphere superiority on the verbal test and a left hemisphere superiority on the nonverbal test, the asymmetries on the two tests being at least as great as for right-handers, and accuracy levels being as high or higher. The direction of lateralization in Group LI was the same as that in Group RN, but the degree of asymmetry was extremely small. Further, the variation in performance accuracy was extreme in this group, some performing unusually well on the verbal test, but poorly on the nonverbal test, some showing the reverse, and some performing poorly on both tests.

Our data could leave no doubt that an inverted hand posture indicates an ipsilateral relationship between the dominant hand and language hemipshere, and that the ipsilateral organization is associated with partial bilateralization of verbal functions, nonverbal functions, or both. Further, people in this group tended to be cognitive specialists, as indicated by their test performance, often being unusually good on one or the other test, but not both. Those performing poorly on both tests may represent the subgroup of pathological left-handers, a not unlikely possibility in view of neurological findings showing that such individuals usually have their major language areas localized to the left side of the brain.

Partial bilateralization of function, as seen in our LI subjects, does not seem to be homologous with cerebral symmetry in other mammals. In nonhuman animals, it must be presumed that there is a single genetic program for hemispheric organization that is expressed bilaterally and symmetrically during development, with each symmetric hemisphere being partially competent in the types of functions differentially served by the two sides of the human brain. In man, it must be presumed that two different genetic programs are available for organizing the hemipsheres during development that, in the majority, are selectively expressed in separate sides of the brain. In LI individuals, it appears that one of the two organizing programs is expressed, to an unusual degree, in both sides of the brain. Thus, in some of these people, one hemisphere is fully organized for verbal functions, while the other is partially organized for verbal and partially for nonverbal processes, and in others, it is the nonverbal functions that are bilateralized. The former manifest quite superior verbal capacities, and deficient visuo-spatial abilities; the latter manifest the reverse.

In a social species, it is not difficult to imagine how frequency-dependent selection could serve to maintain a polymorphism in cerebral laterality patterns that have direct effects on the structure of cognitive abilities. Biologically based differences in cognitive skills and propensities would be likely to serve stability of social organization. While a human group could hardly survive were all its members disposed toward and skilled in a single endeavor, it could only benefit if a few showed special abilities not common in the group as a whole. Those with strongly lateralized brains may be cognitive generalists and competent to fill most social roles, while those with either bilateralized verbal or nonverbal functions, though possibly excluded from certain activities, may be especially able and valuable for others. Data gathered over the last several years show that left-handers are greatly overrepresented in some occupations and underrepresented in others.

Reid, in her doctoral dissertation, has discovered that the <u>direction</u> of cerebral asymmetry has cognitive consequences. Earlier researchers had found that cognitive maturation in children must reach a certain level before hemispheric asymmetries can be revealed in complex tasks (infants, obviously, cannot display a left hemisphere superiority in distinguishing meanings of words), and that in girls, the left hemisphere matured more rapidly, while in boys, the right hemisphere had the developmental advantage. Thus, the left hemisphere advantage in reporting dichotically presented words appears earlier in girls than boys, while the right hemisphere advantage in discriminating nonsense shapes appears earlier in boys than girls. These differences in hemispheric maturation rate are also revealed in standardized cognitive tests, where boys surpass girls on tests of spatial reasoning and girls surpass boys on verbal tests.

Reid tested children of ages 5 and 8 on a test of temporal pattern discrimination (specialized to the verbal hemisphere) and a test of nonsense shape discrimination. Children were classified by sex, handedness, and hand posture during writing. By age 8, all children displayed asymmetries on both tests and in the same direction as we had observed in adults, the children in Group LN being mirror images of those in Group RN, and those in Group LI being weakly lateralized in the same direction as right-handers. At age 5, children manifested an asymmetry on only one of the two tests, opposite for the two sexes within each handedness and hand posture group. Children in Groups RN and LI showed the sex-typical patterns, verbal/left hemisphere asymmetries for girls and spatial/right hemisphere asymmetries for boys, with girls performing better on standardized cognitive tests of verbal function and boys on standardized cognitive tests of spatial function. In Group LN, the central question was whether the right, verbal hemisphere of girls and the left, spatial hemisphere of boys would display asymmetric advantages, or whether the sex-related hemispheric maturation differences would be <u>side-of-brain dependent</u> and <u>function independent</u>. It was this latter that Reid found: the left, <u>spatial</u> hemipshere of girls and the right, <u>verbal</u> hemisphere of boys manifested asymmetric advantages. Further, girls in this group performed best on the standardized spatial test and boys best on the standardized verbal test.

The Reid results show that, independently of its functional specialty, it is always the female left and the male right hemisphere that has the developmental advantage, and that the cognitive patterns shown by children reflect the differing maturation rates of the two hemispheres. What Reid has established is that reversal of the typical direction of lateralization does <u>not</u> produce mirror-symmetric functional organizations. Conceivably, the directional asymmetry in the majority served sex-role differentiation in our prehistoric past, while the minority with reversed asymmetry may have provided a cognitive variation valuable to social organization. Clearly, for such speculations to have any scientific status, we shall need a great deal more research regarding the cognitive structures, occupational choices, and social roles played by individuals who vary in sex, direction of cerebral asymmetry, degree of lateralization, and extent of verbal versus spatial bilateralization of function, both in advanced technological societies, as well as in primitive, neolithic cultures.   When observations relevant to these issues become available, we are likely to gain valuable insights regarding human neurological and social evolution and how they are related.

Until then, I'll leave you with my personal guesses:

Some of us like molecules, and some, the lambda phage.
For others, very simple cells are really all the rage.
Some prefer an embryo; to some it's fairly plain
That nothing's so intriguing as the human and his brain.
It's delightful that we differ, that there's lots of variation,
For our talent altogether makes for social integration.
And if the rarest talents are offered strong protection,
Then our many variations could be due to pure selection.
Should the frequency of genotypes and fitnesses observed
Have a bit of correlation, variation is preserved.

Selected Readings

Levy, J. and M. Reid.   Variations in cerebral organization as a function of handedness, hand posture in writing, and sex.   Journal of Experimental Psychology: General **107**:114-144 (1978).

Levy, J. and C. Trevarthen.   Perceptual, semantic and phonetic aspects of elementary language processes in split-brain patients.   Brain **100**:105-118 (1977).

Levy, J., C. Trevarthen and R. W. Sperry.   Perception of bilateral chimeric figures following hemispheric deconnection.   Brain **95**:61-78 (1972).

Sperry, R. W.   Lateral specialization in the surgically separated hemispheres.   In: The Neurosciences: Third Study Program.   F. O. Schmitt and F. G. Worden, eds.   The MIT Press: Cambridge, Mass. (1974).

Sperry, R. W., M. S. Gazzaniga and J. E. Bogen.   Interhemispheric relationships: The neocortical commissures; syndromes of hemisphere disconnection.   In: Handbook of Clinical Neurology, Vol. IV.   P. J. Vinken and G. W.   Bruyn, eds.   North-Holland: Amsterdam (1969).

NAVIGATION BY HONEYBEES

James L. Gould

Department of Biology
Princeton University

I would like to introduce you to ethology--a field that is relatively new, and, I think, exciting--and to illustrate the sorts of questions ethologists ask and the techniques we use to answer them. The field itself was established by European naturalists--birdwatchers with Ph.D.'s in zoology, mostly--who enjoyed spending their time out in the field watching animals behave and wondering how they managed to seem so smart despite small brains and often numerous legs. In time, these biological birdwatchers gave themselves a name, probably to avoid being confused with psychologists, and gave that name a definition. Ethology, they decided, is the study of an animal in its world, with the goal of understanding the mechanisms that underlie the animal's behavior. As a field, it owes its existence to two fundamental discoveries--that animals live in unique sensory worlds, irrevocably separate from ours; and that they are to a large extent robots preprogrammed by their genes.

The first of these discoveries came around 1915, when Karl von Frisch began wondering why flowers were colorful. He reasoned that color must be present not to please our aesthetic senses, but to attract bees and other pollinators. The unanimous view at the time, however, was that animals in general, and insects in particular, were sensory cripples--deaf, dumb, and virtually blind. To unseat this notion, Frisch designed an elegant test. He trained bees to visit a watch glass full of sugar water, and kept the glass on a colored card. After a time, Frisch set out an array of cards and empty dishes. One card had the same color as the training card, while all the rest were of varying shades of grey. Frisch reasoned that if bees lacked color vision, they should confuse the colored card with at least one of the grey ones. When he used a red card they were confused, indicating that they are blind to red. For all other colors, however, the bees were not fooled, and thus proved that they have color vision. The really interesting part of the story began when Frisch discovered that the bees could distinguish between cards of identical shades of grey which had been made by different companies. After excluding odor as a cue, Frisch found that the cards reflected different amounts of ultraviolet (UV) light, and then proved that bees see UV as a separate color to which we are blind. Frisch asked why bees should want to see in the UV, and quickly discovered that flowers viewed in the UV are transformed into great bulls' eyes.

Frisch went on to emphasize the lesson by discovering that bees can also perceive polarized light to which we are, again, blind. His pioneering work inspired the discovery of several otherwise unimaginable sensory systems in animals: infrared detectors in night-hunting snakes, ultrasonic sonar in dolphins and bats, infrasonic hearing in birds, and magnetic field sensitivity in a variety of animals. Doubtless, other systems are still to be discovered. The lesson is a melancholy one: We are blind to our own blindnesses, and must not try to read our own disabilities into the rest of the animal kingdom.

The other great discovery--that animals are very much like robots--emerged in the 1930's from the studies of Konrad Lorenz and Niko Tinbergen. They found that very simple stimuli could trigger very complicated behavior. One dramatic example is known as the "egg-rolling response." Many ground-nesting birds will roll an egg back into their nests if it happens to tumble out. At first glance this seems a straightforward and sensible piece of behavior, and suggests some amount of intelligence on the bird's part. Lorenz and Tinbergen, however, thought it looked suspiciously stereotyped. Upon investigation, they found the entire sequence to be the mindless working of a machine. The geese they used as experimental subjects would roll in anything even vaguely round, including beer cans and volleyballs--objects which, once in the nest, would be recognized by the bird as what they were and disdainfully ejected. More bizarre still, the object could be removed any time after the bird had reached for it, and the now-imaginary egg would nevertheless be rolled back gently into the nest. The lesson here is that animals can be preprogrammed by their genes to do complicated and apparently intelligent things.

These two stories have not been entirely without a point. The behavior I want to discuss deals with the programming behind a very complex piece of navigation that depends upon cues to which we are completely blind. It is a part of the system by which forager honey bees get from their hives to a food source and back again without getting lost. This may not sound like much of a trick, until you look at things from the bee's point of view. For one thing, bees must fly enormous distances in search of food, often 1 to 10 km. This is the equivalent of our venturing forth on a journey of 60 to 600 miles. Then too, bees do not see all that well. Their compound eyes can just resolve a daisy as a flower from 6 inches away, and so their navigational strategy, like that of human explorers, is to use celestial cues as a compass to get them close enough to their goal to begin a visual search. Landmarks are also used, but they must be very conspicuous to be of any great value.

The key to understanding how bees navigate came with Frisch's discovery of the "dance language." A returning forager encodes the distance and direction of a food source into a dance that potential recruits attend, decode, and use to locate the food for themselves. The distance is specified by the <u>duration</u> of the waggle run, while the direction is indicated by its <u>orientation</u> (Figure 1).

FIGURE 1 — *Waggle dance. Returning foragers perform a dance which communicates the distance and direction of the food to other bees.*

Outside, with a view of the sky, foragers aim their dances directly at the food, while inside, in the darkness of the hive, where the dances are performed on vertical surfaces, the angle between the waggle run and vertical corresponds to the angle between the sun and the food. These arbitrary conventions of defining "up" as the direction of the sun and each waggle as corresponding to so many meters allow the dance language to work. And what are the components of language but a collection of arbitrary conventions shared within a culture?

Frisch soon realized that the dances indicated something about what information bees were using on their journeys, and how they were using it. For example, the sun was clearly their primary cue. But the sun is a problematical landmark. It moves from east to west at an ever-changing rate, and Frisch discovered that bees somehow compensate for that motion in both their flights and their dances. On windy days, bees must aim themselves into crosswinds in order to maintain a course. Hence, the angle of their bodies with respect to the sun is altered. When they dance, however, they indicate the true angle between the sun and the goal rather than the one they had to face on their way there.

Similarly, Frisch found that when forced to fly to the food by a circuitous route, foragers nevertheless indicated the true compass direction in their dances. They had somehow integrated the separate legs of the flight. More remarkable still, Frisch discovered that bees navigate perfectly well when the sun disappears behind clouds, trees, or the horizon. Somehow they are able to infer its position. He was able to show that this ability depended on polarized UV light coming from the sky. For example, when he hid the sun from bees dancing outdoors, the dances continued as before, but when he interposed a UV-absorbing filter, they became disoriented. Later, he found that a UV-passing polarizer would reorient the dances.

As an ethologist, I am interested in the programming that underlies each of these remarkable feats of navigation and communication. My colleague, Michael Brines of the Rockefeller University, and I set out to understand how honey bees use UV polarized light to locate the sun--information that must then be fed into the more general navigation program. When we started, the general guess among ethologists was that bees use the geometrical patterns of polarized light in the sky predicted by simple Rayleigh scattering, and that those patterns are best in the UV (Figure 2). Our backgrounds in physics, however, told us that neither of these suppositions was very likely.

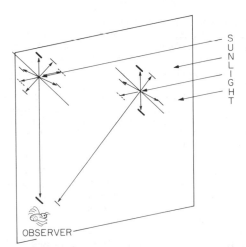

FIGURE 2 — Skylight polarization is created when incoming sunlight is scattered by the atmosphere. The direction and degree of polarization thus generated depend on the scattering angle in a very regular way. As illustrated here for two points in the sky, the polarization is always perpendicular to the plane containing the sun, the point of scattering, and the observer; and the degree of polarization is greatest for scatterings of 90° and least for 0° and 180°. The result, in theory at least, is the beautifully symmetrical patterns depicted in Figure 3.

Rayleigh scattering predicts that when the unpolarized light from the sun is scattered by air molecules, a net polarization will be induced perpendicular to the plane defined by the incoming and outgoing rays; and that the degree of polarization will be determined by the scattering angle, being greatest for 90°. The result ought to be a predictable pattern symmetrical with respect to the sun (Figure 3). The orientation and degree of polarization at any point in the sky should specify the distance and direction of the sun. However, this whole story depends on the atmosphere being isotropic and optically inactive, and on light from the sun being scattered only once before reaching us. This is certainly never the case in nature, and when relatively crude measurements in visible light were made several decades ago, the patterns did not look very good. However, the apology was generally made that since the bees use UV, the patterns must be well-aligned there.

Dr. Brines and I began by actually measuring the patterns of polarized light available to bees. We used a narrow-band skylight polarimeter of our own

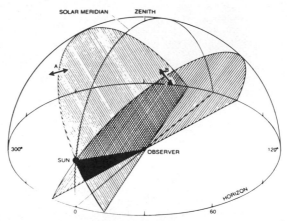

FIGURE 3 — *Polarization pattern. Simple Rayleigh scattering predicts a symmetrical pattern of polarized light in the sky.*

design and construction, and a computer that aimed it and took the data. The resulting maps of the sky indicate that the patterns are no better in the UV on either clear or overcast days, and that the degree of polarization is, as any survivor of Feynman physics would expect, somewhat lower. Hence, at first glance it looks as if the UV is not a particularly inspired choice of wavelength. In looking at our data again, however, we think we finally <u>have</u> discovered a decisive advantage for the UV. We found that light must be at least 10% polarized for bees to orient to it. On partly cloudy days, the clouds reflect enormous amounts of unpolarized sunlight at visible wavelengths--which is why we can see them--and this reflection swamps out the small amounts of polarized light created by atmospheric scattering between the cloud and the observer. On the other hand, as anyone knows who has gotten sunburned on an overcast day, clouds pass rather than reflect most UV light. As a result, the clouds are nearly transparent in the UV, and so the degree of polarization remains above the perceptual threshold for bees only at this wavelength.

Having now some idea of what bees can actually see in the sky, the next question must be how the information is used. The way we ask the bees is by training some to an artificial food source, and then forcing them to dance on a horizontal surface in the dark. Deprived of the "up" rule, they are disoriented. We then offer them various sorts of artificial sky patterns whose wavelength distribution, elevation, angular size, brightness, and polarization orientation and degree are under our control. Their dances tell us whether or not the cues are meaningful, and if they are, how they are being used. For example, the first problem for a bee dancing outdoors with a restricted view of the sky is to decide what to orient her dance to. If the sun is visible, it takes precedence over sky polarization; but how does a bee decide whether it is seeing the sun? This seems trivial to us, but the visual resolution of the bee's eye is such that it cannot perceive the sun directly. Indeed, eight suns could fit into the field of view of one ommatidium.

In fact, bees have a rule by which they use the size, color, and polarization of what they can see to decide (Figure 4). The rule is mostly arbitrary. The elevation of the patch and its absolute brightness are ignored; and while the sun has no polarization, a visual angle of $1\frac{1}{2}°$, and (depending on conditions) a UV content of less than 8%, the rule will cause dancers to identify a 100% polarized, 15°, 20% UV spot confidently as the sun. The rule corresponds to physical reality only in that the sun actually will be identified as such, and communication fails only for stimulus combinations that do not exist in the natural sky. Otherwise, the rule seems quite arbitrary--a case of willful self-deception. Isolated pieces of blue sky or cloud seen through the trees by a dancer will often be taken to be the sun, and the dances are therefore oriented inappropriately. Of course, since the attending bees use the same rule to orient their interpretation of the dance, no misunderstanding takes place. In fact, our calculations indicate that for most of the parts of the sky that the bees mistakenly identify as sun, dances and recruit flights will be more precisely oriented than if the stimulus had been correctly identified.

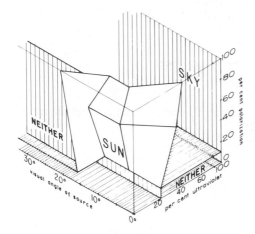

FIGURE 4 — *Sky/sun rule. In deciding whether they are seeing the sun or part of the sky, bees use a rule which takes into account the degree of polarization, the angular size, and the relative amount of UV light in the source.*

We found two more arbitrary conventions when we began concentrating on how bees use the parts of sky they actually call sky to locate the sun. The major difficulty they face is that at any particular elevation, identical polarization angles may be found in many pairs of places in the sky. Hence, when they see a pattern, the bees may have to decide which one of a pair it is before they can locate the sun. Now the simplest way to make this distinction would be to use the percent polarization as an index of how far away from the sun the spot is. In fact, our observations indicate that bees ignore percent polarization, and it is a good thing they do since our direct measurements indicate that this theoretically ideal relationship is never seen in the natural sky. Three other possible methods for making this crucial distinction remain: The bee might dance to both interpretations, or she might pick one on her own arbitrarily, or she might have a rule for deciding. The first two methods would result in half of the recruits being dispatched in the wrong direction, while the latter, if all the bees agreed on it, would result in no misunderstanding. Bees do have such a rule. It dictates that the pattern is always to be interpreted as the one that exists furthest from the sun. Hence, though technically wrong much of the time, this arbitrary convention allows the language to work.

In looking over our sky maps, however, we noticed that at almost any elevation there are two discrete points with the same polarization angle that are the same distance from the sun. Since the "furthest" rule must fail, we wondered how bees dealt with the situation. When we presented these patterns, the bees inevitably fell back on another rule: The patch of sky was always taken to be the one that is located to the right of the sun. Again, although this right-hand rule results in the stimulus being mistakenly identified half of the time, the mistakes are systematic and no misunderstanding ensues.

These three newly discovered rules, like the up-is-the-sun rule and the meters/waggle convention (a rule, by the way, that varies from 5 to 75 m/wag between races), provide the arbitrary linguistic conventions that are essential to social communication. As such, bees possess the only abstract, symbolic language in nature that we know of other than our own.

At this point, however, having fought our way through the rules, we still must ask how bees use polarized light to locate the sun. Three hypotheses come to mind. The first is that bees perform spherical geometry: They draw a great circle perpendicular to the polarization vector and note its two intersections with the sun's elevation (Figure 5). Then they apply the rules to decide which intersection to use.

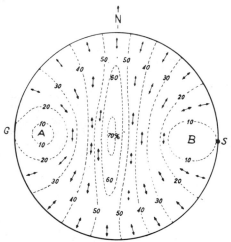

FIGURE 5 — *Great circle solution. In theory, the sun should lie on the great circle passing perpendicularly through the polarization vector, and be located at one of the circle's two intersections with the sun's elevation. From "Polarized-Light Navigation by Insects" by Rudiger Wehner. Copyright © 1976 by Scientific American, Inc. All rights reserved.*

A second possibility is that bees are solving the problem analytically: They might calculate the relative azimuth of the sun using a set of differential equations. A group of European scientists has even derived a workable set empirically. Since two solutions would usually be generated by the procedure, the new rules might then be used to decide which one should be chosen.

A third possibility, originally suggested by Frisch and subsequently laughed to scorn as impossibly simpleminded, is that bees might simply "photograph" the sky, and then try to match the "picture" taken in the field with the cues available overhead during their dance. Where two places in the picture match the cue, the rules would be used to select the "correct" one. Around Princeton, the photograph hypothesis used to be known as the "joke theory," but now our data indicate that Frisch may, as usual, have guessed right. The decisive experiment has yet to be done, but preliminary evidence is suggestive: Under our experimental conditions, dances are less well-oriented on cloudy days.

How does the poor orientation of dances on cloudy days argue for the photograph theory? Under either the geometrical or analytical hypotheses, the accuracy of the forager dances should depend on the quality of the cues available during the dance and the accuracy with which the forager knows the angle between the sun and the food. Since we were showing the bees the same patterns that gave precise orientation on sunny days, the fault would have to lie in their knowledge of the sun's position. In fact, the bees knew perfectly well where the sun was since, unlike us, they can see it through the clouds in UV light; and when offered an artificial sun in the hive, the dances instantly became precisely oriented. Frisch's unfashionable photograph theory, on the other hand, can account for the data perfectly well: Clouds reduce the degree of polarization reaching the bees during their flights, so the dancer had to work with a very fuzzy picture when she tried to match it with the artificial cues. Hence, the dances were poorly aimed on cloudy days, but well-oriented on sunny ones. We will be testing this notion more rigorously in the future.

I began by saying that the strategy of ethology is to discover the mechanisms that underlie behavior. After the basic, overt behavior of an animal has been described--the fun part--there are at least three levels of mechanisms at which the dissection must be done. The most superficial questions are those that ask what the logic of the network is--the programs and sub-routines that produce the behavior we see. The second level of questions must be directed at learning how nerves are connected together to produce this programming and generate behavior. At the third level, we must ask how the information in the genes is used to build and connect those nerve cells together in the first place.

These four levels--genes, nerves, programming, and behavior--form a causal chain that evolution has forged and left for us to decipher. Except for programming, we have a pretty good idea of the general principles at work in each of these levels. The real challenge in learning how behavior is created comes in trying to understand how each level gives rise to the next link up. The other great challenge for ethology--molecular or otherwise--is to apply its insights to our own species: to discover how the unseen hands of evolution are directing or predisposing our behavior now that we have, at least overtly, left our hunter-gatherer heritage so very far behind.

Selected Readings

Brines, M. L. and J. L. Gould. Bees have rules. Science, in press (1979).
Tinbergen, N. The Study of Instinct. Oxford University Press: Oxford (1972).
von Frisch, K. The Dance-Language and Orientation of Bees. Harvard University Press: Cambridge, Mass. (1967).

PLASTICITY OF TRANSMITTER MECHANISMS
IN SYMPATHETIC NEURONS

Edwin J. Furshpan

Department of Neurobiology
Harvard Medical School, Cambridge

One of the central ideas, practically a dogma, in neurobiology is that neurons only make synaptic connections with a specific subset of other neurons and that this pattern of specific connections (the "wiring diagram") underlies the patterns of information processing that take place in the nervous system. Another, perhaps minor, dogma is that one can characterize neurons by the neurotransmitter or the neurohormone that they synthesize and secrete. Neurons have a transmitter "flavor"; one talks about cholinergic neurons, or adrenergic neurons, tryptaminergic neurons, and so on. Peptidergic neurons are another important class.

These two notions of connection specificity and transmitter specificity come together when one considers how the nervous system is formed. Not only do neurons have to make contacts with the proper targets but the components of the synaptic apparatus obviously have to be appropriately matched. The transmitter released by the presynaptic endings of the synapse have to be appropriate for the receptors in postsynaptic cells. Because of the exacting specificities of postsynaptic receptors, if there is a mismatch, the synapse won't function at all. Furthermore, in order for the function of the synapse to be appropriate--for example, excitation instead of inhibition--the combination of receptor and transmitter has to be appropriate.

The vertebrate heart provides a familiar example illustrating these notions; as we all know, the heart receives a dual innervation from the autonomic nervous system, one set of inputs from the sympathetic nervous system, and another from the parasympathetic nervous system. For the moment, I will talk only about the final neurons in these two pathways--the ganglionic sympathetic neuron, and the ganglionic parasympathetic neuron. We know that the ganglionic sympathetic neuron, when it is activated, releases norepinephrine onto the heart and causes its beat to increase in frequency and strength; and conversely, when the parasympathetic ganglionic neuron is activated it releases acetylcholine onto the heart and slows down the beat. Because of its reflex connections, the sympathetic neuron becomes active in situations where it is appropriate for the heart to become more active. Clearly it is important that this neuron secretes norepinephrine; if it were to secrete acetylcholine, the resulting inhibitory action would be inappropriate and possibly disastrous.

This raises the question of what determines the transmitter flavor of autonomic neurons. The rule isn't simply that all sympathetic neurons become adrenergic and all parasympathetic neurons become cholinergic. There is quite good evidence that some sympathetic neurons, perhaps a few percent, are cholinergic. This isn't a random error in development; the cholinergic sympathetic neurons innervate particular target organs (e.g., sweat glands). In our experiments we were surprised to find that this determination of transmitter flavor can take place quite late in the development of the sympathetic neruons. It takes place in postmitotic neurons taken from animals after birth--that is, at a fairly late stage.

The experimental system we work with consists of isolated sympathetic neurons dissociated from rat sympathetic ganglia and put into cell culture. I will show you how we do that in just a moment, but first I would like to mention a number of people who have been involved in working on this project over the years. Among the physiologists, in addition to David Potter and myself, very importantly have been Paul O'Lague, Peter MacLeish, and Colin Nurse. We were joined by a biochemist early on--Dennis Bray--who devised some of the methods that were important in getting the system to work. We were later joined by Paul Patterson, who has headed up the biochemical effort on these experiments ever since. The chief collaborators with Paul on the experiments I am going to describe here were Dick Mains, Linda Chun, Lou Reichardt, Pat Walicke, and Michel Weber. I will mention each of their contributions as we go along. And finally, the electron microscopy was done first by Philippa Claude and more recently by Story Landis.

The neurons we use are taken from the largest of the sympathetic ganglia, the superior cervical ganglia. We initially chose to work with mammalian sympathetic neurons for two reasons. First, these neurons have a diversity of targets in the body. They innervate heart muscle, a variety of smooth muscles, some gland cells, brown fat, and--in primates at least--the liver parenchyma, and perhaps the parenchyma of some other tissues. Second, sympathetic neurons have a specific growth factor that is very well characterized, the nerve growth factor (NGF). Levi-Montalcini and her colleagues have shown that this factor not only stimulates the growth of neurons but is required for their survival. Beyond some minimal amount which they need for survival, the growth of the neurons is related to the concentration of NGF. Aside from the inherent interest in NGF, it seemed useful to have one of the required growth factors in order to simplify our medium. With NGF, we do not need to add complicated mixtures like chick-embryo extract, which is often otherwise necessary.

The method for isolating the neurons was worked out by Dennis Bray: the ganglia are decapsulated and teased apart with fine forceps. The remarkable result of this is that clouds of neurons, singly or in clumps, are liberated into the medium. The large clumps are removed, and the single neurons and small clusters are plated into cell-culture dishes. Several weeks later, some assay, physiological, biochemical, or anatomical, is performed. The ganglion contains non-neuronal cells as well as neurons. However, the number of non-neuronal cells in the initial suspension is very small, and one can either inhibit their division--for example, by gamma irradiation--or one can kill them off with a metabolic poison like cytosine arabinoside. By one of these methods, one can grow the neurons in more or less pure culture. Alternatively, one can use a medium which is permissive for the proliferation of ganglionic non-neuronal cells, or one can add non-neuronal cells of one's own choosing, to get cultures in which the neurons are growing in the presence of many non-neuronal cells.

During the first day or so in culture, the neurons begin to send out processes (axons). At the end of each process is the characteristic growth cone. The growth cones split from time to time, and the two parts go off in different directions, each spinning out an axon behind it. Each of these growth cones may split, in turn, and in this way a rather extensive axonal tree is formed. After several weeks in culture, with several thousand neurons in the dish doing this, a very dense network of axons is formed.

The left side of Figure 1 shows a culture, several weeks old, in which the ganglionic non-neuronal cells have not been permitted to divide and almost all of the cellular material is neuronal. One can see the cell bodies of eight neurons; the rest of the dense mass of tissue consists of axons and some dendritic processes. The right side of Figure 1 shows a mixed culture, one grown in a medium permissive for the multiplication of the non-neuronal cells. One can see the cell bodies of five neurons and their dendritic processes; the remainder is a carpet of ganglionic non-neuronal cells on which the neurons lie and through which the axonal processes are interlaced. (The axons are obscured by the ganglionic non-neuronal cells.)

Our original intention was to confront the neurons in culture with various

143

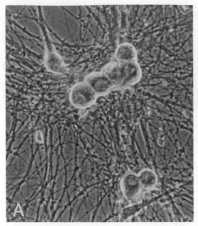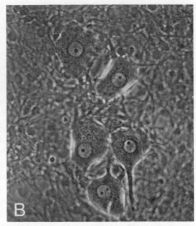

FIGURE 1 — *The appearance of sympathetic neurons from superior cervical ganglia of rats after several weeks in culture; phase-contrast micrographs of small fields in mass cultures that each contain about 1,000 neurons. The field shown in A is from a culture grown in a (bicarbonate-free) medium non-permissive for the proliferation of non-neuronal cells; only neuronal cell bodies and processes are seen. The field shown in B is from a culture grown in permissive medium; the initially few ganglionic non-neuronal cells have multiplied to form a continuous carpet on which the neurons lie. The neurons do not divide in either medium.*

potential target cells (heart cells, gland cells, etc.). However, we were soon distracted from this attempt by the observation that the neurons frequently made functional synapses with each other. These synapses were all excitatory. That is, a nerve impulse evoked in one neuron would depolarize the other neurons to which it was functionally connected. In many cases, the synaptic effect was suprathreshold, causing the receiving neurons also to fire impulses which, in turn, would evoke a barrage of impulses in most or all of the other neurons in the culture. This prevalence of excitatory synapses among sympathetic neurons was unexpected. As indicated above, most adult sympathetic neurons (all but a few percent) are adrenergic, and norepinephrine and other adrenergic agents were known to have inhibitory effects on sympathetic neurons. Indeed, we then found, by use of routine pharmacological tests, that these synapses were not adrenergic, but were cholinergic. The prevalence of these effects indicated that many or most of the neurons in such cultures (in the most favorable cases, more than 75%) were cholinergic.

This prevalence of cholinergic neurons was not found in all cultures. It eventually became clear that it occurred only when the neurons were grown in the presence of a large number of non-neuronal cells (Figure 1; right side). In neuron-alone cultures, usually less than 1% of the neurons, and never more than 3%, came up cholinergic. This wasn't because the neurons grown by themselves were sick or simply failing to make or release neurotransmitter. Experiments to be described next showed that the neurons in neuron-alone cultures are predominantly adrenergic.

Richard Mains and Paul Patterson began their biochemical studies of these cultures by examining the ability of the neurons to synthesize a number of neurotransmitters from the appropriate labeled precursors. They used neuron-

144

alone cultures, so that all of the transmitter synthesized could be ascribed to the neurons. The cultures were selective in their synthetic abilities; they failed to make detectable amounts of several transmitters--γ-aminobutyric acid, serotonin, octopamine, and histamine. Some of the cultures, and only those more than 3-4 weeks old, did make small amounts of acetylcholine. However, all of the cultures synthesized (from labeled tyrosine) the catecholamines, norepinephrine, and dopamine (dopamine is the immediate precursor to norepinephrine in these neurons); and this ability increased dramatically as the neurons continued to develop in culture.

The greatest increase was during the second week. By the third week, the neurons were very adrenergic indeed, incorporating a significant fraction of the total tyrosine they took up into catecholamines. The steepness of the rise during the second week was much greater than the concomitant rise in the ability to synthesize protein or lipids or RNA. So it seems to be a specific differentiation which the neurons are undergoing.

There are other indications that when the neurons are grown by themselves, they are perfectly good adrenergic neurons. Louis Reichardt and Patterson found that these neurons would take up labeled norepinephrine by means of a high affinity uptake mechanism and, when depolarized by the addition of potassium ions to the medium, would release this norepinephrine. The release mechanism had the conventional property of being calcium-dependent. In summary, these neurons can synthesize, take up, and release norepinephrine just the way normal sympathetic noradrenergic neurons do.

The next obvious question was whether, with the biochemical assay, neurons growing in the presence of non-neuronal cells would come up making substantial amounts of acetylcholine. The answer Patterson and Chun found was a definite yes. If they allowed the non-neuronal cells to proliferate or added extra non-neuronal cells, they found that after a while the neurons made very substantial amounts of acetylcholine.

Another question was whether this effect requires direct contact between the two types of cells, or whether it can be transferred by way of the medium. To test this, Patterson and Chun did a "conditioned medium" experiment. They grew neuron-alone cultures of the type that normally become adrenergic. They also grew non-neuronal cells in another vessel, and fed the neuron cultures every couple of days with medium (conditioned medium, or CM) from the non-neuronal cells. In fact, they examined the effect of various proportions of CM to fresh medium; they never used more than about 65% CM so that the neurons would always have some access to fresh nutrients. The neurons were fed with the different concentrations of CM for 2-3 weeks before being assayed.

The effects which they observed in this conditioned-medium experiment were dramatic. As the proportion of CM to fresh medium was increased, the neurons made more and more acetylcholine and less and less norepinephrine. To get some sense of how large a change occurred in the synthetic ability of the neurons, one can compare the ratio of acetylcholine to norepinephrine synthesized in the different conditions. It was found that this ratio increased about a thousandfold from neurons grown without CM (100% fresh medium) to neurons grown in 62% CM.

The number of neurons surviving in the cultures was independent of the concentration of CM and did not change much with time in individual cultures. This argues against a simple selection mechanism: one in which there are two separate populations of neurons--one predetermined to be cholinergic, the other predetermined to be adrenergic--and, depending on the concentration of CM, varying proportions of these two populations survive. There is a variant of this hypothesis in which the neurons that are selected against don't die, but fail to make any neurotransmitter--so-called "silent" neurons. This possibility is not ruled out by the experiment just described since, simply by counting, one can't distinguish silent from synthesizing neurons. However, experiments on single, isolated neurons, that I'll describe later, make this variant on the selection hypothesis also unlikely. It looks very much as if the non-neuronal cells (and CM) are having an inductive, rather than a selective, effect--that is, almost all of the neurons in a culture survive and synthesize transmitter, but are changed over in the presence of CM from an initial adrenergic state to a final cholinergic state.

There are several lines of evidence that suggest that these neurons have already begun to develop along the adrenergic pathway by the time we take them (from newborn animals) and put them into culture. At an early stage during the development of sympathetic ganglia in vivo, some days before birth, the neurons display the aldehyde-induced fluorescence characteristic of catecholamine-containing neurons. Also, if the neurons are examined in the electron microscope (after fixation in $KMnO_4$) during their first few days in culture, it is found that almost all of their synaptic vesicles have the dense cores characteristic of adrenergic neurons. In fact, Story Landis has found that the initial growth cones sent out by the neurons during their first day or so in culture contain these characteristic adrenergic vesicles; this is true even when the neurons have been exposed to CM during this day or so in culture. Our hypothesis, then, is that these neurons receive an early instruction, in vivo, to become adrenergic; in culture, this signal can be countermanded by another instruction from non-neuronal cells to become cholinergic.

In mass cultures of this sort, in which several thousand neurons are synthesizing varying amounts of both norepinephrine and acetylcholine, it is natural to wonder what the individual neurons are doing. During this period in culture, do two classes of neurons develop, one of which is adrenergic and one of which is cholinergic, or are there individual neurons that can synthesize and/or release both transmitters? To approach this question, it is not good enough to dissect out a single neuron cell body from a mass culture and assay it, because the cell body will have on it presynaptic endings from other neurons and these will confuse the assay. The most straightforward way to go is to grow single neurons, in isolation, in microcultures. We have used two methods for this purpose. Reichardt and Patterson plated neurons at low density into multiwell plates and after a suitable culture period assayed transmitter-synthesis in those wells containing a single neuron. For our electrophysiological assays, we made microcultures consisting of many small islands of heart cells growing on a plastic coverslip; again, some time after neurons were plated at low density, those islands to which a single neuron had adhered were studied. An example of such a microculture is shown in Figure 2. The primary function of the heart cells in these microcultures is to provide a simple and sensitive bioassay for the neurotransmitter(s) released by the neuron. The heart cells continue to beat in culture and retain their conventional sensitivities to the autonomic neurotransmitters: their beat is inhibited by acetylcholine and is accelerated by norepinephrine.

To make the assays, we insert one microelectrode into a heart cell in order to monitor the rate of beating, and insert another microelectrode into the neuron in order to stimulate it electrically and, at the same time, to record its electrical activity. In our physiological experiments, we have so far only examined microcultures in a rather restricted age range, 2-3 weeks. In such cultures, we have found three classes of neurons: those that were conventionally cholinergic, those that were conventionally adrenergic, and those that were in between.

Cholinergic neurons were identified in two ways. First, stimulation of a neuron in this class caused slowing or, more usually, cessation of the beating of the heart cells (see Figure 3A); that this inhibition was mediated by ACh was verified by the use of atropine, a specific blocker of ACh-mediated effects on the heart. In addition, such neurons usually formed cholinergic synapses on themselves, so that an action potential directly evoked in the neuron gave rise

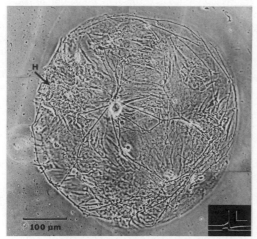

FIGURE 2 — A microculture containing a single neuron on an island of heart cells; phase-contrast micrograph. The numerous processes of the neuron can be seen extending over the surface of the heart cells. The neuron had been in culture for 19 days. The arrow at H indicates a clump of heart cells from which electrical recordings were made; the action potential in the inset (lower right) was recorded from the neuron which was adrenergic (see Figure 3B).

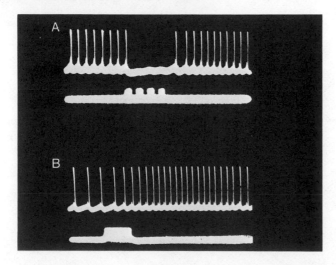

FIGURE 3 — The effects of stimulating single neurons on the beating of heart myocytes in microculture. In each pair of recordings, taken from two different cultures, the upper trace shows the action potentials recorded from the heart cells; the lower trace, at low gain, shows the action potentials of the neurons and indicates the periods during which trains of stimuli were delivered to the neurons. A: the characteristic inhibition of the heart cells evoked by stimulating a cholinergic neuron; this inhibition was blocked by atropine (not shown). B: the characteristic excitation of the heart cells evoked by stimulating an adrenergic neuron; the excitation was blocked by propranolol (not shown). The duration of the entire sweep in each record was about 15 sec.

to excitatory synaptic potentials in that same neuron; in some cases these synaptic potentials were large enough to evoke another action potential and in this way the neuron would re-excite itself several times. There is anatomical evidence for the presence of such self-synapses, or autapses, in the central nervous system. In these microcultures, since only a single neuron is present, one obtains the most direct evidence that a neuron can make functional synapses on itself. The cholinergic receptors on these neurons have long been known to differ from those on heart and smooth muscle, and are specifically blocked by hexamethonium or curare. Thus, the presence of hexamethonium-sensitive autaptic responses in this class of neurons provided additional confirmation that they were cholinergic.

The second class of neurons, those that were conventionally adrenergic, were recognized by their excitatory effect on the heart. When a neuron of this class was stimulated, it caused depolarization of the heart myocytes and an increase in their rate of beating (see Figure 3B). The abolition of this effect by propranolol, one of the commonly used specific antagonists of adrenergic effects in the heart, provided additional evidence.

The third, and most common, class of neurons were unconventional. They caused a combination of the effects described for the first two classes, thus blurring the usually sharp distinction between cholinergic and adrenergic neurons. Stimulation of such neurons evoked curare-sensitive (cholinergic) excitation of themselves, atropine-sensitive (cholinergic) inhibition of the heart cells, followed by propranolol-sensitive (adrenergic) excitation of the heart cells (see Figure 4). It was clear that the cholinergic and adrenergic effects were both caused by the same neuron, since only a single neuron was present in the microculture. We refer to neurons of this type as dual-function neurons.

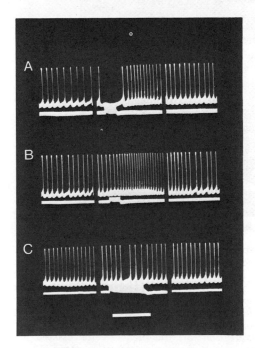

FIGURE 4 — *The effect on the heart cell of a dual-function neuron. The three sets of recordings (upper traces, from the heart cells; lower traces, from the neuron) were all taken from the same single-neuron microculture. The records in A illustrate the characteristic sequence of inhibition followed by excitation of the heart cells caused by stimulation of a dual-function neuron. The inhibition was blocked by 0.1 μM propranolol to the atropine solution (C). Time mark: 10 sec. The breaks in the records are the gaps between successive oscilloscope sweeps. The neuron had been in culture for 13 days.*

When Story Landis examined dual-function neurons in the electron microscope, she found that their synaptic endings were of mixed appearance: they contained the clear (electronlucent) vesicles characteristic of cholinergic endings as well as the dense-cored vesicles characteristic of adrenergic endings. In the cases examined so far, the clear vesicles have predominated.

The presence of this dual-function state is consistent with our hypothesis that the neurons have already begun to develop along the adrenergic pathway at the time we harvest them but can be induced in culture to make a switch in transmitter choice and to develop along the cholinergic pathway. According to this idea, the dual-function state is a transient reflection of the changeover; the neurons have begun to synthesize and secrete acetylcholine before they have stopped secreting norepinephrine.

Most of these electrophysiological experiments were made on neurons that had been in culture, in the presence of heart cells, for 2-3 weeks. This is the period during which Patterson and Chun find the most striking changes in transmitter metabolism. This may account for the fact that the majority of the neurons we examined during this period were dual-function. According to our hypothesis, the neurons should eventually complete their differentiation along the cholinergic pathway, or along the adrenergic pathway if they failed to switch, and then dual-function neurons should no longer be present. We have only just begun to examine older neurons with the physiological assay, but Reichardt and Patterson have made observations with biochemical methods consistent with this prediction. They assayed the ability of neurons to synthesize acetylcholine and norepinephrine, and found that after 4-5 weeks in microculture, isolated neurons generally synthesized only one of the two transmitters; only 3 of 105 were found to make detectable amounts of both. They found that the proportion of cholinergic to adrenergic neurons increased with the addition of more conditioned medium or heart cells to the microcultures. Also consistent with the changeover hypothesis was their finding that almost all of the neurons made at least one transmitter; only 2 of 105 neurons were "silent" with respect to transmitter synthesis. This is inconsistent with the "silent neuron" version of the selection hypothesis described above.

The most direct test of the changeover hypothesis would be to assay the transmitter flavor of the same neuron at different times during its development in culture. We have found that this is technically feasible. The electrophysiological assay for transmitter flavor can be made several times on the same neuron at intervals of a week or so. Preliminary observations from such experiments indicate that a shift in transmitter profile does, in fact, occur.

These observations raise several obvious questions and I would like to finish by addressing two of them. First, what is the nature of the conditioned-medium "factor": is there, in fact, a single factor; do the non-neuronal cells add something to the medium, or do they alter or remove components already present (e.g., in the serum)? One possibility that initially seemed attractive is that the effective substance is nerve-growth factor (NGF). Chun and Patterson have indeed found that beyond a certain minimal amount required for neuronal survival, the more NGF added to the culture, the more norepinephrine the cells synthesize. However, contrary to the predictions of this hypothesis, acetylcholine synthesis is also enhanced by NGF when the neurons are grown under cholinergic conditions (with CM). Chun and Patterson found that the amounts of the two transmitters that are synthesized increase in parallel as more NGF is added, and the ratio of these amounts remains constant. NGF affects survival, growth, and extent of differentiation of the neurons but not the choice of transmitter. In contrast, CM factor has only a minimal effect on survival and growth, but affects the choice of pathway along which differentiation occurs. NGF is permissive; CM factor is instructive.

If CM factor is not NGF, what is it? Michel Weber and Patterson have been working to purify the factor and have achieved a considerable increase in its specific activity. A serious difficulty impeding the progress of these studies is the lack of a quick assay for the factor. It now takes 10 days to 2 weeks to determine whether the neurons have become adrenergic or cholinergic. However, Weber and Patterson have found that the factor behaves like a large molecule with an apparent molecular weight of about 50,000, and its activity is destroyed by periodate oxidation.

I'll finish by considering the question of what prevents most sympathetic neurons from becoming cholinergic during normal development in vivo, where there is clearly no shortage of non-neuronal cells. There are a number of differences between the situations of the neurons in culture and in the body. The endocrine environment in culture, in the absence of hormones (except those in the small amount of serum added) is strikingly abnormal. The neurons in culture do not become well ensheathed by the satellite cells that enclose them in vivo; perhaps the satellite cells normally protect the neurons from CM factor. Another obvious deficit is the absence of the neural input normally supplied from the spinal cord by the preganglionic neurons.

In order to imitate the excitatory influence conveyed by this neural input, Patricia Walicke, Robert Campenot, and Patterson have used several techniques to excite or depolarize the neurons in culture. They have stimulated the neurons electrically, added the drug veratridine which holds open sodium channels in the neuronal membrane, or depolarized the cells by increasing the concentration of potassium ions in the medium. They have found that any of these procedures, imposed chronically, has a remarkable effect on the neurons: it prevents the cholinergic-inducing effect of conditioned medium. That is, in the presence of electrical activity or passive depolarization, the neurons continue to develop along the adrenergic pathway although exposed to a concentration of CM that would otherwise make them predominantly cholinergic. In addition, they have found that calcium ions play an essential part in this effect. If the entry of Ca^{++} into the neurons is prevented by the use of standard Ca^{++}-channel blocking agents (D-600, Mg^{++}), the neurons retain their responsiveness to CM (i.e., they become predominantly cholinergic) even though they have been chronically depolarized. Conversely, elevated concentrations of Ca^{++} (or of Ba^{++}), in the absence of Ca^{++}-channel blockers, enhance the effect of depolarization.

These observations suggest a speculation about the mechanisms that determine which sympathetic neurons become adrenergic and which become cholinergic during normal development in vivo: perhaps those neurons that first receive an effective presynaptic input from the spinal cord, and are thus brought into activity early in their development, are made insensitive to the non-neuronal cell factor(s) and therefore continue to develop along the adrenergic pathway; those neurons that receive their input later, or have inputs that become active later, remain sensitive during a critical period to the factor(s) that now switches their development onto the cholinergic pathway. An attractive aspect of this idea is that it suggests a relationship between the neurospecificity mechanisms that lead neurons to make synaptic connections only with an appropriate subset of other neurons (or muscles, etc) and the inductive mechanisms that determine the appropriate transmitter flavors of neurons.

Selected Readings

Furshpan, E. J., P. M. MacLeish, P. H. O'Lague and D. D. Potter. Chemical transmission between rat sympathetic neurons and cardiac myocytes developing in microcultures: Evidence for cholinergic, adrenergic, and dual-function neurons. Proceedings of the National Academy of Sciences, Washington, 73:4225-4229 (1976).

O'Lague, P. H., D. D. Potter and E. J. Furshpan. Studies on rat sympathetic neurons developing in cell culture. III. Cholinergic transmission. Developmental Biology 67:424-443 (1978).

Patterson, P. H. Environmental determination of autonomic transmitter functions. Annual Review of Neuroscience 1:1-17 (1978).

THE LEECH EMBRYO

Gunther S. Stent

Department of Molecular Biology
University of California, Berkeley

C. O. Whitman and T. H. Morgan

The year 1978 is memorable not only because we celebrate the Golden Jubilee of the Caltech Biology Division, but also because it marks the centenary of the publication of Charles Otis Whitman's seminal paper on the embryology of leeches. In that paper, Whitman presented for the first time the notion that the early cells of an embryo, or blastomeres, are already differentiated with respect to the role that each will play in the future course of the animal's development. More importantly, in his 1878 paper Whitman presented the first analysis of developmental cell lineage and described the exact succession of cleavages by which the germ layers of the leech embryo arise from the fertilized egg.

Ten years later, in 1888, Whitman became the first director of the Woods Hole Marine Biological Laboratory and took on the role of a kind of late-Victorian Max Delbruck of American embryology. At Woods Hole, Whitman figured as the inspirational leader of a group of younger biologists, whose names became much better known than that of their leader. Among these disciples of Whitman were Edmund B. Wilson, Edwin Conklin, Frank Lillie, and for this occasion most importantly, Thomas Hunt Morgan.

All of these men were then carrying out developmental cell lineage analyses at Woods Hole on leeches, other annelid worms, and mollusks. As this audience is, of course, aware, Morgan had started out as an embryologist and eventually turned to his genetic studies with Drosophila mainly because he thought that it would be necessary to fathom the mechanism of heredity before the really deep biological problem in want of explanation, namely ontogeny, could be understood.

Morgan never lost sight of that problem and in his later years expressed his disappointment that so little progress had been made in answering the question that had engaged him in his youth, namely why, as he phrased it, an egg gives rise to an ostrich rather than to a hippopotamus. And, I imagine, Morgan knew that the reply "Why, in that egg there are ostrich genes and not hippopotamus genes" has no standing as an embryological explanation. So, it would have been good for some speaker to ask at this Symposium commemorating Morgan's founding of the Biology Division: "Whatever happened to embryology these past 50 years?" Unfortunately, I am not the person to provide such a retrospective. But just so that embryology does not go entirely unmentioned in these proceedings, I propose to close this Symposium with an old-fashioned, almost entirely descriptive account that continues on from where Whitman left leech embryology at the turn of this century.

This account is the very antithesis of the only kind of talk that Max Delbruck tolerated when I first took my orders at Caltech 30 years ago. But I feel justified all the same in presenting to you this purely descriptive material without any grand take-home lessons, because of my firm conviction that Whitman's leech embryo--unaccountably neglected by developmental biologists since the turn of the century--offers tremendous promise for future work.

Leeches

Leeches constitute the Order Hirudinea in the Phylum Annelida of

151

segmented worms. Among the features that set leeches apart from other annelid worms is the fixed number of 32 segments that make up their body. Of these 32 segments, the frontmost four form the specialized structures of the head, including paired eyes on the dorsal surface and a front sucker on the ventral surface. The rearmost seven segments form the specialized structures of the tail, including an anus and a large rear sucker. The intervening 21 midbody, or abdominal, segments form a highly stereotyped iteration of visceral organs, such as circulatory vessels, kidneys and gut. Leeches are hermaphrodites and carry male and female genital orifices on the ventral midline of their 5th and 6th abdominal segments, respectively. The external surface of each body segment is subdivided into transverse annuli, of which one contains circumferentially distributed segmental sensory organs, or sensillae. The segmental body wall consists of several different sets of muscles. The circular muscles form one of these sets; their contraction reduces the diameter of the body and thus, since its total volume filled with an incompressible fluid must remain constant, causes elongation of the body. Deep in the circular muscles lie their antagonists, the longitudinal muscles. Contraction of the longitudinal muscles shortens the body, and consequently, increases its diameter.

The two modes of locomotion of the leech--swimming and walking--are effected by means of such changes in body shape. For walking, the leech attaches its rear sucker to a solid substrate and elongates its body in a plane parallel to that of the substrate by contraction of the circular muscles. The front sucker is then attached to the substrate, the rear sucker is released, and the body is shortened by contraction of the longitudinal and relaxation of the circular muscles, resulting in a head stand. Now relaxation of the dorsal and further contraction of the ventral longitudinal muscles in midbody segments produces a hairpin bend and brings down the rear sucker into contact with the substrate next to the anterior sucker. The rear sucker is then attached, the front sucker is released and raised off the substrate by contraction of the dorsal and relaxation of the ventral longitudinal muscles in its front segments. Finally, the now S-shaped body is elongated by contraction of the circular and relaxation of the longitudinal muscles, for the completion of one walking step.

For swimming, the leech generates a wave of rearward moving crests and troughs along its elongated body. The crests and troughs are produced by the rhythmic contractions of the longitudinal muscles of the ventral and dorsal segmental body wall, respectively. The period of the alternating ventral-dorsal segmental contraction cycle is of the order of 1 sec, and the rearward movement of the crests and troughs is the result of a progressive front-to-rear phase delay of the contractile cycles of successive segments. The rearward travel of the troughs and crests pushes against the fluid medium and thus propels the leech forward.

The leech nervous system

The central nervous system of the leech reflects the segmental body plan. It consists of a chain of 32 interconnected ganglia, which form the ventral nerve cord. Each ganglion contains the cell bodies of about 175 pairs of bilaterally symmetrical neurons, as well as a small number of unpaired neurons. The ganglion is connected to the body wall and to the internal organs via two bilateral pairs of segmental nerves. It is connected to its neighboring ganglia of the nerve cord via bilaterally paired nerve tracts, the connectives. In the region of the head, the frontmost four segmental ganglia are fused to form a large head ganglion. In the region of the tail, the rearmost seven ganglia are fused to form a large tail ganglion.

As extensive neurological studies with the medicinal leech, Hirudo medicinalis, have shown, the anatomy of the remaining non-fused 21 ganglia of the abdominal segments is sufficiently stereotyped from segment to segment,

152

and sufficiently invariant from individual to individual, that a large fraction of their neurons can be reproducibly identified. These neurons can be penetrated with microelectrodes to secure recordings of the electrical activity of individual cells.

Microelectrodes can be used also for the intracellular injection of histologic stains, such as horseradish peroxidase, in order to reveal the anatomical details of individual neurons. By means of these techniques, it has been possible to ascertain during the past dozen years or so the modalities, such as light, touch or pressure, and the receptive fields of many sensory neurons, as well as the target muscles of many motor neurons. Moreover, some interneurons have been identified whose axons lie entirely within the central nervous system and which carry out such higher level functions as the integration of sensory information and the generation of complex movements.

Most importantly, it has been possible to establish the manner of intra-segmental as well as intersegmental synaptic connection of these sensory, motor and interneurons, and thus account not only for some simple acts of reflexive behavior, such as the shortening of the body in response to touch and the raising of the annuli in the skin in response to pain, but also for some moderately complex integrated muscle movements, such as the heartbeat.

The most complex behavioral routine thus far given an account in terms of identified neurons and their connections is the swimming movement. Our own work during the past few years has shown that the contractile rhythm of the segmental longitudinal muscles responsible for swimming is generated by an ensemble of a dozen bilateral pairs of rhythmically active motor neurons present in each segmental ganglion of the ventral nerve cord. Their activity rhythm is imposed on the motor neurons by four bilateral pairs of oscillatory interneurons present in every segmental ganglion. These interneurons form a rhythmically active network, the central swim oscillator.

Thanks to this fairly detailed knowledge of the functional elements of the nervous system of the leech, it is possible to pose quite specific and clearly focused questions regarding its development. In particular one may ask how in the course of embryonic development the particular identified sensory neurons, motor neurons, and interneurons of the segmental ganglia arise and how they become specifically interconnected to form the identified neuronal circuitry. That is to say, the leech nervous system provides a clearly defined conceptual endpoint for the study of ontogenetic processes.

Glossiphonid leeches

Unfortunately, the neurologically well-characterized medicinal leech Hirudo is not particularly suitable for embryological studies. It is difficult to breed in the laboratory. Moreover, the egg of Hirudo (just as that of the other members of the leech family of Gnatobdellae) is of microscopic size. It is laid in a hard-cased cocoon filled with a nutrient fluid on which the embryo depends for nourishment during its development. Thus, in Hirudo the early, and in many respects crucial, stages of embryogenesis are not readily accessible to direct observation and experimental manipulation. By contrast, the egg of leeches of the family Glossiphonae on which Whitman had carried out his pioneering studies, is much larger and rich in yolk. It is laid in a membranous sac, from which it can be removed, and, when maintained in dilute saline, will develop into a juvenile leech, drawing entirely on its own yolk for nourishment.

Recently we have brought two glossiphonid leech species suitable for embryological studies into continuous laboratory culture, to provide us with a constant supply of freshly laid eggs. One of these species is Helobdella triserialis, a carnivorous leech living in the United States, feeding on aquatic snails. The other species is Haementaria ghilianii, a blood-sucking leech living in French Guiana feeding on mammals.

Helobdella and Haementaria lay eggs of diameters of about 0.5 mm and 2.5 mm, respectively.   Embryonic development begins immediately upon deposition of the eggs, and after a week or two, a juvenile leech has arisen that differs from the adult form mainly by its smaller size.   Subsequent, post-embryonic growth of the juvenile leech represents mainly an increase in cell size, rather than an increase in cell number.   The adult form of Helobdella reaches a length of only about 1 cm, which renders it unfavorable for neurophysiological experimentation.   However, its short egg-to-egg generation time (6 weeks, compared to 6 months for Haementaria) and the ease of breeding it make it an attractive species for pilot experiments.

The adult Haementaria, by contrast, reaches a length of about 50 cm (making it the largest known leech), with an experimentally readily accessible nervous system.   Fortunately, the principal anatomical and physiological features of the nervous system of Haementaria resemble those of Hirudo, although there exist also some obvious neurological differences between the two species.

The following account of the embryonic development is based on observations made both on Helobdella and Haementaria, as well as on a third species, Theromyzon rude. (The last is a blood-sucking leech feeding on aquatic birds. Unfortunately, Theromyzon is difficult to breed in the laboratory. For embryological studies, its freshly deposited eggs can be collected in local ponds during the breeding season.) The major features of the embryonic development of glossiphonid leeches are very similar, and no distinction will be made here between the three species. This account is based on work carried out in my laboratory during the past three years by Roy T. Sawyer, Juan Fernandez, David A. Weisblat, Andrew Kramer, and Seth Blair.

Formation of blastomeres and teloblasts

The bulk of the glossiphonid egg cytoplasm consists of yellow yolk, with the remainder being composed of a colorless teloplasm. The onset of embryonic development is marked by a concentration of the teloplasm at the animal and vegetal poles of the egg. The first cleavage occurs about half a day after the egg is laid. That first cleavage is meridional and divides the egg into two cells of unequal size: a smaller cell AB and a larger cell CD. Most of the teloplasm passes into CD. Another half-day later, the second cleavage, also meridional, has given rise to four cells, the blastomeres A, B, C, and D. Of these, cell D is the largest and receives most of the teloplasm. Blastomere D divides again into two cells of nearly equal size, cells D1 and D2. This is followed by cleavage of cell D2 in a plane orthogonal to that of the previous cleavage of cell D, and yields the right and left teloblasts Mr and Ml, with Mr lying directly under cell D1.   According to Whitman, the M teloblast pair is the precursor of mesodermal tissues. Cell D1 cleaves to yield the bilateral cell pair NOPQr and NOPQl. The plane of this cleavage is meridional and lies in the future midline of the embryo. The NOPQ cell pair cleaves nearly synchronously, to yield two bilateral pairs of daughter cells. The smaller pair is designated as the right and left teloblasts Nr and Nl, and the larger as OPQr and OPQl. Next, the OPQ cell pair cleaves nearly synchronously to yield the smaller teloblast pair Qr and Ql, and the larger pair, PQr and PQl. Finally, the OP cell pair cleaves to yield the fourth and fifth teloblast pairs Pr and Pl, and Or and Ol. The N, O, P, and Q teloblast pairs are the precursors of ectodermal tissues. Between them the five teloblast pairs contain most of the teloplasm of the egg.   According to Whitman, the N teloblast is the precursor of the nervous system.

It should be noted that up to this stage, morphogenesis of the embryo is attributable mainly to the cell cleavage pattern. According to this pattern, the first two orthogonally meridional cleavages that give rise to cells A, B, C, and D are followed by the oblique cleavage of cell D, which in turn is followed by the orthogonally oblique cleavage of cell D2 and the meridional cleavage of cell D1. The progeny cells of D1 then undergo a series of orthogonally oblique cleavages, whose planes show mirror symmetry across the future midline of the embryo.

Formation of germinal bands

As soon as it has been formed, each teloblast begins to carry out a series of iterated, unequal divisions. These divisions produce a column of small, disk-shaped stem cells, to which most of the teloplasm, but little of the yolk, of the teloblast is eventually parcelled out. The five pairs of stem cell columns produced by the five teloblast pairs grow in a frontward direction and merge on either side of the midline to form a prominent symmetric pair of cell ridges, the germinal bands. With the ongoing production of more and more stem cells, right and left germinal bands advance rostrolaterally and come to line the right and left lateral edges of the future dorsal surface of the embryo. Right and left germinal bands converge at the front end of the dorsal surface, site of the future head. After reaching the lateral edge, the now crescent-shaped germinal bands continue their circumferential movement and enter the future ventral surface. Eventually right and left germinal bands meet on the ventral midline, where they coalesce. Coalescence begins at the rostral end--the future head-- and continues zipper-like in a rearward direction. When coalescence has gone to completion, the coalesced germinal bands lie in the ventral midline. In the course of movement and coalescence, the stem cells of the germinal bands begin to produce the precursor cells of the adult tissues. This cell multi-plication results in a gradual broadening of the germinal bands, to give rise to the germinal plate.

Segmentation

The earliest sign of incipient segmentation of the embryo appears soon after the beginning of germinal band coalescence. At that time, the first formation of somites is manifest at the front end of the embryo, as a longitudinal fragmentation of the germinal plate tissue into a series of tissue blocks. Somite formation then progresses in a rearward direction. When germinal band coalescence is about two-thirds complete, the next sign of incipient segmentation is provided by the agglomeration in a front-to-rear sequence of paired cell masses on either side of the germinal plate midline. Each of these cell mass pairs represents the primordium of a ganglion of the ventral nerve cord. The first four frontmost primordia to be formed are the precursors of the head ganglion. The next 21 primordia to be formed are the precursors of the midbody, or abdominal ganglia, and the last and rearmost seven primordia eventually form the tail ganglion. Meanwhile, formation of the incipient 32 body segments has gone to completion, with the boundary between successive germinal plate tissue blocks being provided by a transverse septum.

Development of the nerve cord

The first sign of the site of formation of a ganglionic primordium is given by a local coalescence on the ventral midline of the right and left medial stem cell rows produced by the Nl and Nr teloblast pairs. Successive sites of stem cell coalescence are separated by regions of non-coalescence, which mark the interganglionic space to be occupied by the future nerve cord connective. At the sites of coalescence, the stem cell descendants of the N teloblasts begin to proliferate in a lateral direction, to produce on either side of the ventral midline a fan-shaped sheet of neuroblasts in the plane of the germinal plate. This sheet eventually folds over to produce the globular ganglion. This

155

morphogenetic folding process is attended by growth of connective tissue that eventually encloses the ganglion in a sheath and separates the nerve cell bodies into anterior and posterior lateral pairs of cell packets, and into anterior and posterior ventromedial cell packets. Differentiation of the neuroblasts into neurons, as manifest by the outgrowth of axons from the cell body into the interior of the primordium, begins during the folding process. By the time folding is complete, formation of an incipient neuropile in the interior of the globular structure is under way. Lateral growth of axons beyond the confines of the developing ganglion produces the two bilateral pairs of segmental nerves. Of these, the pair of anterior nerves grows within the intersegmental septum, near which the ganglion is located.

Meanwhile, formation of the intersegmental connective begins by axon outgrowth into the next anterior and next posterior ganglion. The frontmost four ganglia of the cord, being the first to develop, remain together and fuse to form the head ganglion. But from the fifth ganglion rearward, successive ganglia of the cord move further and further apart, attended by elongation of the connective and its axons that already connect the embryonic ganglia. Finally, the rearmost seven ganglia, being the last to develop, also remain in close proximity and fuse to form the tail ganglion. Since the embryonic nervous system lies on the ventral surface of the embryo, it is possible to take intracellular recordings from the neurons practically as soon as the ganglia are morphologically intact.

Development of body wall and gut

The completion of the body wall of the leech proceeds by a bilaterally circumferential expansion of the germinal plate tissue from ventral midline over the surface provided by the A, B, and C blastomeres. This expansion of the germinal plate proceeds via further division of the stem cells of the germinal bands. Thus, circular muscles arise from the growth of oblong myoblasts oriented perpendicularly to the longitudinal axis. By contrast, the underlying longitudinal muscle fibers appear to be deposited on either side of an anatomically distinct, longitudinally oriented growing zone. This growing zone progresses circumferentially and, at the time that the expanding right and left halves of the germinal plate finally meet on the dorsal midline, the growing zone lies on the lateral edge of the now nearly mature embryo.

The endodermal gut tissue is derived from the A, B, and C blastomeres. By the time that formation of the ganglionic primordia is complete, the gut appears as a long cylinder (filled with yolk provided by the blastomeres) extending from the 5th to the 20th somite. Segmentation of the gut then proceeds in a front-to-rear direction, by way of a series of annular constrictions of the yolk-filled cylinder. Each constriction is in register with an intersegmental septum, giving rise to about 15 paired gut expansions, or caeca, in register with an abdominal body segment. Upon completion of the gut segmentation process, the embryo has taken on the general body form characteristic of the adult leech.

A tracer technique for cell lineage analysis

The lines of descent from the uncleaved egg of individual macromeres and of teloblasts and their stem cell columns described in the foregoing was established by direct microscopic examination of developing embryos. But after those early developmental stages, the method of direct observation becomes too cumbersome for following the further fate of individual cells. Hence, in order to allow us to ascertain in more detail the pedigrees of particular cells of the mature leech body, we developed a new cell tracer technique.

This technique uses horseradish peroxidase (HRP) as an intracellular

156

tracer, injected through a micropipette into identified cells of early embryos. After HRP injection, embryonic development is allowed to progress to a later stage, at which time the distribution pattern of HRP within the tissues is visualized by staining for its presence (the staining is produced by the action of the HRP on a reagent consisting of hydrogen peroxide and benzidine, as a result of which any cell containing HRP takes on an intense dark color). This method proved feasible because in <u>Helobdella</u> embryos, three essential conditions are met: 1) after injection of HRP, embryonic development continues normally; 2) the injected HRP remains catalytically active and is not diluted too much in the developing embryo; and 3) HRP does not pass through the gap junctions that interconnect the cytoplasms of the embryonic cells.

An example of the use of this method is provided by an experiment in which the N teloblast of an early embryo was injected with HRP before production of its stem cell column had begun. Development was then allowed to proceed for six more days, by which time formation of the ventral nerve cord, with its head and tail ganglionic masses and the intervening chain of 21 segmental ganglia on the ventral midline of the germinal plate, was complete. After staining for the presence of HRP, it was seen that the half of the ventral nerve cord ipsilateral to the injected N teloblast was darkly colored. This result confirms Whitman's indirect inference that the N teloblast and the stem cell column to which it gives rise is the major precursor of the ipsilateral half of the leech nervous system.

If, in a similar experiment, the N teloblast is injected with HRP at a somewhat later stage, after production of the stem cell column had already begun, then it is found that the posterior, but not the anterior, part of the ipsilateral ventral nerve cord is stained. This result shows that the rear part of the nervous system develops from the last stem cells produced by the N teloblast. Since in such an experiment the boundaries between stained rear and unstained front portions of the cord, and between stained ipsilateral and unstained contralateral halves of the ganglia, is very sharp, it would appear that during development of the nerve cord there occurs little migration of cells either longitudinally or across the ventral midline of the embryo.

A second example of the use of this method is provided by injection of the M teloblast of an early embryo. After allowing development of the embryo for six more days, and staining for the presence of HRP, it is seen that many mesodermal tissues and organs ipsilateral to the injected M teloblast take on the dark color. However, in this case the ventral nerve cord is unstained. This finding provides direct confirmation for Whitman's indirect inference that the M teloblast is the precursor of the mesoderm.

Cell ablation

Thanks to the highly determinate nature of leech embryogenesis, it is possible also to establish developmental cell lineages by observing the eventual effects of ablating a particular cell of the early embryo. For this purpose, we have adapted a specific cell-killing technique recently developed by I. Parnas and D. Bolling at Stanford University. According to this technique, identified cells of the early embryo are killed by injecting them via a micropipette with the potent proteolytic enzyme pronase.

An example of the use of this method is provided by an experiment in which the left mesodermal teloblast M1 was injected with pronase at a stage where formation of its stem cell column was partially complete. Development of the embryo was then allowed to proceed to the nearly mature juvenile stage. The result of this experiment was that whereas the front part of the animal developed quite normally, of the rear part, only the right half was normal. The left rear part (including the left half of the rear sucker), was completely disorganized and lacked any recognizable organization of its tissues, and, in particular, lacked any segmental character. Hence, it can be concluded that the absence of the mesodermal stem cell column in the posterior part of the left germinal band prevented not only the formation of mesodermal tissues but also averted the normal morphogenesis of ectodermal tissues in the left rear body. This suggests that in the leech, as in other animals, the mesoderm plays an important role in the organization of the ectoderm.

Ontogeny of movement

A striking feature of leech embryos is that their active movement begins at an early stage of development and then evolves as a stereotyped sequence of motor acts. This sequence consists of the progress from simple, irregular contractions of individual muscle fibers to complex, concatenated movement patterns. Moreover, some of the simple movements of early developmental stages clearly form part of some more complex movements displayed at later stages. And these more complex movements are, in turn, evidently components of locomotory routines of juvenile and adult leeches. Some of these embryonic movements may fulfill some physiological function at the stage at which they occur, but others may represent practice of adult behavior, in order to guide the appropriate, functional connections among the components of the developing central nervous system.

The first visible movement is a front-to-rear peristaltic wave which begins at the stage when the embryo is still enclosed in the egg membrane and the germinal plate covers only about one-third of its ventral surface. By then the germinal plate has a fully-formed ventral nerve cord with ganglia and connectives. Moreover, circular muscle fibers already envelop the whole embryo, coursing circumferentially in an orientation perpendicular to the longitudinal axis. And it is the coordinated contraction-relaxation rhythm of these circular muscles that produces the peristaltic wave.

Peristalsis appears to function in hatching. Each wave pushes the head against the egg membrane, and it seems as if this repeated pushing eventually ruptures the membrane and allows escape of the embryo, head first, from its casing. Soon after hatching, the embryo begins to carry out new types of movements, while the peristaltic waves diminish in frequency and eventually disappear altogether.

By now the germinal plate has expanded to cover most of the ventral surface of the embryo, and contains longitudinal muscle fibers that are already innervated functionally by motor neurons of the ventral nerve cord. The new types of movements consist of an alternating shortening of the body due to a concurrent contraction of the longitudinal muscles and an elongation of the body due to a concurrent contraction of the circular muscles. Shortening and elongation occur at irregular intervals and both eventually reach an average frequency of about one movement per minute, with each contraction lasting from about 5 to 25 seconds.

Two new types of movements appear once expansion of the germinal plate has finally covered the dorsal surface of the embryo. These are ventrally concave or dorsally concave bendings, attributable to contraction of the longitudinal muscles on the ventral or dorsal surface, respectively, each occurring about once a minute. Eventually, by the time that segmentation of the gut and formation of the caeca nears completion, these spontaneous movements are combined into what appears to be the precursor of the walking movement. In one such primordial "walking step," the entire embryo first elongates by concurrent contraction of circular muscles. It then shortens by concurrent contraction of the longitudinal and relaxation of the circular muscles. Next the embryo takes on a hairpin shape, by further contraction of the ventral and relaxation of the dorsal longitudinal muscles in the rear. This is followed by bending of the embryo into an S-shape by contraction of the dorsal

158

and relaxation of the ventral longitudinal muscles in the front. Finally, the S-shaped embryo elongates by contraction of the circular and relaxation of the longitudinal muscles.

All that is still needed for maturation of this behavioral routine into actual walking is attachment of the rear sucker followed by release of the front sucker after the hairpin bend and attachment of the front sucker followed by release of the rear sucker after shortening. The embryos gain the capacity to use their suckers in this manner within the next few days and are now walking juvenile leeches. Shortly thereafter, they also gain the ability to carry out the swimming movement.

Future projects

As is evident from this account, we have not yet advanced very far towards providing an answer to Morgan's (paraphrased) problem of why there develops a leech from the egg rather than a hippopotamus. But I think several more narrowly-focused questions relevant to this problem can at least be posed at this time. First, we may ask how the precise morphogenetic succession of alternating cell cleavage planes of the early leech embryo is actually determined. This determination probably reflects the replication and predetermined positioning of centrioles, whose orientation appears to govern the cleavage planes in mitosis. And the replication of centrioles is likely to involve novel, hitherto unknown, processes by which complex structures built of protein molecules can serve as the direct template for the assembly of their replicas or mirror images.

Second, we may inquire into the nature of the teloplasm that is passed on only to the D blastomere and eventually parcelled out by the teloblasts to their stem cell descendants. The teloplasm clearly plays some critical role in development, and by transferring it via microinjection from cell to cell, it should be possible to gain some insight into its function and chemical identity.

Third, we may seek to fathom the mechanism of the segmentation process, by which there arises the fragmentation of the germinal plate tissue (presumably of the mesodermal cell descendants of the M teloblast) into a series of tissue blocks, or primordial somites. One possibility would be, of course, that in the column of mesodermal stem cells and their descendants there develops an alternating pattern of surface features, say feature A and feature B, such that cells carrying the same feature adhere to each other and cells carrying a different feature repel each other. But how the embryo arranges to divide the germinal plate into exactly 32 blocks of self-adhering mesodermal tissue--never 31 or 33--is not so easy to explain. It is unlikely that the number 32 results from the initial presence of a given amount of some substance, say the teloplasm, which, when divided up 32 times, is just enough for the formation of one primordial somite. For in that case the fluctuations both in the initial amount of that substance and in its partitioning during the segmentation process would be expected to be sufficiently great that individual leeches with a number of segments other than 32 should be frequently encountered. It seems more likely therefore that the precise segment number is determined by some geometric process involving five quantal elements or steps (e.g., successive activation of five different genes), resulting in $2^5 = 32$ primordial somites.

Finally, we may ask whether the development of the motor neurons and their connections that underlie the gradual perfection of the walking movement of the leech requires practice, or whether the adult locomotory behavior arises autonomously in the absence of any functional feedback. It should be possible to obtain an answer to that question by following the maturation of motor routines in embryos which have been immobilized during their development by exposure to paralyzing pharmacologic agents.

In closing I can affirm only my stouthearted belief that by the time of the

Biology Division's Diamond Jubilee in 2003, the leech will have become both the phage and the fruit fly of embryology.

Selected Readings

Mann, K. H. Leeches (Hirudinea). Pergamon: New York (1962).
Nicholls, J. G. and D. Van Essen. Scientific American **230**:38 (1974).
Stent, G. S., W. B. Kristan Jr., W. O. Friesen, C. A. Ort, M. Poon and R. L. Calabrese. Science **200**:1348 (1978).
Weisblat, D. A., R. T. Sawyer and G. S. Stent. Science **202**:1295 (1978).
Whitman, C. O. Quarterly Journal of Microscopical Science **18**:215 (1978).

AUTHOR INDEX

Ambesi, S., 103
Ames, B., 2, 3, 21-31, 98
Ananiev, E. V., 90
Anderson, T., 41
Arber, W., 70
Armelin, H., 101
Atkinson, J., 112
Beadle, G. W., 2, 32, 79, 81-87, 98
Benacerraf, B., 109, 110
Benjamin, T., 6
Benzer, S., 41, 61, 121-122
Berg, P., 6
Berget, P., 66
Bertani, J., 41
Bingham, P., 88-92
Blair, S., 154
Blum, A., 28
Bogen, J., 124
Bolling, D., 157
Bonner, J., 39, 121
Borsook, H., 3
Boveri, T., 24, 33
Bray, D., 143
Brenner, S., 42, 61
Brill, W. J., 93
Brines, M., 137
Britten, R., 90
Broca, P., 123
Bruce, B., 29
Buonassisi, V., 100
Campbell, J., 39
Campenot, R., 150
Casida, J. E., 28
Chun, L., 143, 145, 149
Clark, J., 101
Claude, P., 143
Cohen, M., 121-122
Conklin, E., 151
Coon, H., 103
Crick, F. H. C., 69
Cullen, S., 111
David, C., 111
Davidson, E., 39, 90
Davidson, N., 90

Davis, R., 73
Dax, M., 123
Delbrück, M., 4, 39-43, 61, 72, 75, 97, 98, 122, 151
Demerec, M., 72
Dobzhansky, T., 80
Doermann, G., 41
Donahue, V., 27
Donelson, J. F., 88
Dulbecco, R., 1, 2, 3, 4-10, 11, 13, 18, 41, 101
Dunsmuir, P., 88-92
Dworkin, M., 115
Edgar, R., 41, 42, 61, 62, 64, 65, 66, 123
Eiserling, F., 44-45
Ellis, E., 39, 40, 61
Emerson, R. A., 81
Epstein, D., 41, 42, 61
Erikson, R., 7
Fernandez, J., 154
Feynman, R., 42, 138
Finnegan, D. J., 88
Franklin, N., 41
Frelinger, J., 111
Freund, R., 92
Fried, M., 5
von Frisch, K., 121, 135, 136, 137, 140, 141
Furshpan, E. J., 121, 142-150
Furth, J., 100
Galinat, W., 84
Gazzaniga, M., 124
Georgiev, G. P., 90
Godson, N., 49
Gorer, P., 107
Gospodarowicz, D., 101
Gould, J. L., 121, 135-141
Green, M. M., 88
Grohmann, K., 50
Gutierrez, M., 84
Gvozdev, V. A., 90
Hamilton, C., 131
Harm, H., 41

161

SUBJECT INDEX

Cell culture, 98, 100, 143; conditioned medium (CM), 145; differentiated, 98; enrichment, 100; establishment of differentiated cell lines, 101-103; hormone-supplemented medium and, 101-103; serum-free medium and, 101-103; tissue, 98

Cell-mediated lympholysis reaction, 110

Chimu-Inca, 81

Chondromyces, 114; fruiting body of, 114; slime trail of, 114; swarming of, 114

Chronic myelocytic leukemia, 36

Cigarette smoking condensate as carcinogen, 25, 27; genetic defects, 27

Coeliac disease, 105

Competition hybridization, 76

Complement factors, 112; association with HLA or GPLA and, 112; Bf, 112; C2, 112; C4, 112; evolution of, 112

Complementation: in vitro T4, 64-70

Conditional lethals, 42, 46; T4, 61, 62

Corn (Zea), origin and genetics of, 81-87; chromosome, 81; genetics, 81; hybrids with teosinte, 81, 84, 85; pollen, 85

Cycasin as carcinogen, 26

D locus. See D region of H-2 locus

D region of H-2 locus, 106, 107, 108; cytotoxicity and, 106, 110; evolution of genes of, 108; products of, 108, 113

DDT as carcinogen, 23, 26

Density gradient centrifugation, 42

Dibromochloropropane as carcinogen, 28, 30-31

Dieldrin as carcinogen, 23, 26

Diethylstilbestrol as carcinogen, 25, 24

Differentiated tumor cells, 100; alternate passaging of, 100; isolation of, 100, 101

DNA replication: ØX174, 55-59; semi-conservation, 42; unwinding, 56

Drosophila, 79, 80, 88-92; imaginal discs, 90; melanogaster, 88, 89, 90, 92; polytene chromosome, 88; simulans, 88, 89, 90, 92

Embryology of the leech, 151, 154-159

Enterogenic hormones, 101

Environmental carcinogens, 2, 21-31; Salmonella as assay for, 2, 22, 24

Ethology, 135-141

Ethylene dibromide as carcinogen, 28

Ethylene dichloride as carcinogen: precursor to vinyl chloride, 22

Fat intake and cancer, 29

Fecalase, 26

Fibroblast growth factor, 101, 102

Fibroblast overgrowth in tissue culture, 98, 99; dedifferentiation hypothesis and, 99; selection hypothesis and, 99

Fibronectin, 102

Frameshift mutation, 42

Furylfuramide as carcinogen, 27

Gene system A of myxobacteria, 117

GH₃ cells, 101, 102

Gonadotrophins, 101

Graft rejection, 110

Graft vs. host reaction, 110

Griseofulvin as carcinogen, 26

Gross virus-induced leukemogenesis, 106, 108

Growth hormone, 101

H gene, 108

H-2 system, 104, 106, 107; congenic inbred strains, 106, 107, 109, 111; disease susceptibility and, 106, 113; duplicated loci and, 108

H-2D genes, 110; mutations in, 110

H-2K genes, 110; mutations in, 110

Hair dye as carcinogen, 28

Handedness and brain organization, 129-132; hand posture and cerebral asymmetries, 131-132

Hemisphere specialization, 123-133; benefits of, 131; congenital asymmetries, 130-131; and handedness, 129-132; in monkeys, 131; selective activation, 126; sex-role differentiation, 133; right hemisphere linguistic ability, 127-129

Heteroduplex mapping, 73, 74, 75

Histidine biosynthesis, 2

"Hit and run" hypothesis, 5

HLA system, 104, 105; D region of, 104; disease susceptibility and, 105; DR region of, 104; evolution of genes of, 108; I region of, 104;